UNDERSTANDING AND CALCULATING THE ODDS

Probability Theory Basics and Calculus Guide for Beginners, with Applications in Games of Chance and Everyday Life

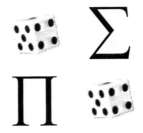

Cătălin Bărboianu

Cătălin Bărboianu
PO Box 1 – 230
Craiova, Romania
cb@infarom.ro
http://probability.infarom.ro

INFAROM Publishing
Pure and applied mathematics
office@infarom.ro
http://books.infarom.ro
http://www.infarom.ro

ISBN 973-87520-1-9

Mathematics subject classification (2000):
00A05, 00A07, 00A08, 00A30, 03A05, 03C95,
60A05, 60A10, 60A99, 65C50, 65C99.

Contents

Introduction

Every one of us uses the word *probable* few times a day in common speech when referring to the possibility of a certain event happening.

We usually say an event is *very probable* or *probable* if there are good chances for that event to occur.

This simplistic dictionary definition of the *probable* quality attached to an event is unanimously accepted in current speech.

Far from standing for a rigorous definition, this enunciation still gives evidence for the quantitative and measurement aspect of the probability concept because the *chance* of an event occurring is represented by figures (percentages).

Meanwhile, these chances are the numerical result of an estimation or calculation process that may start from various hypotheses.

You will see in the following chapters how the proper probability calculus can lead to a variety of numerical results for the same event. These results are a function of the initial information that is taken into account.

In addition, establishing a certain threshold from which the chance of an event occurring attribute to it the quality of being *probable* or *very probable* is a subjective choice.

All these elements create a first view of the relativity of the term *probable* and of the possible errors that can be introduced into the qualitative and quantitative interpretation of probability.

These interpretation errors, as well as that false *certainty* psychologically introduced by the numerical result of measuring an event, turn probability calculus into a somewhat dangerous tool in the hands of persons having little or no elementary mathematical background.

This affirmation is not at all hazardous, because probabilities are frequently the basis of decisions in everyday life.

We estimate, approximate, communicate and compare probabilities daily, sometimes without realizing it, especially to make favorable decisions.

The methods through which we perform these operations could not be rigorous or could even be incorrect, but the need to use probability as criterion in making decisions generally has a precedence.

This could be explained by the fact that human beings automatically refer to statistics in any specific situation, and statistics and probability theory are related.

We usually take a certain action as result of a decision because statistically, that action led to a favorable result in a number of previous cases.

In other words, the probability of getting a favorable result after that action is acceptable.

This decisional behavior belongs to a certain human psychology and the human action is generally not conditioned by additional knowledge.

Although statistics and even probability do not provide any precise information on the result of a respective action, the decision is made intuitively, without reliance on scientific proof that the decision is optimum.

For example, when we say, *"It will very probably rain today,"* the estimation of a big chance of rain results from observing the clouds or the weather forecast.

Most of time it rains when such clouds are in the sky—according to statistics—therefore, it will very probably rain today, too.

In fact, this is an estimation of the probability of rain, even though it contains no figures.

If we take an umbrella, this action is the result of a decision made on the basis of a previous estimation.

We choose to take an umbrella not because we see clouds, but because most of time it rains when the sky is cloudy.

When we say, *"It is equally probable to obtain 1 or 3 after rolling a die,"* we have observed that the die has six sides, and so the number of possible outcomes is six.

Among those possible outcomes, one corresponds to occurrence of 1 and one to occurrence of 3. So, the chances are equal (1/6).

Unlike the first example, the previous estimation of chances resulted from a much more rigorous observation, which led to a numerical result.

Another example—otherwise unwanted—of making a decision based on probability is the following:

Your doctor communicates the stages of evolution of your disease: if you won't have an operation, you have a 70 percent chance of living, and if you'll have the operation, you have a 90 percent chance of a cure, but there is a 20 percent chance that you will die during the operation.

Thus, you are in a moment when you have to make a decision, based on personal criteria and also on communicated figures (their estimation was performed by the doctor according to statistics).

In most cases of probability-based decisions, the person involved performs the estimation or calculus.

Here is a simple example:

You are in a phone booth and you must urgently communicate important information to one of your neighbors (let us say you left your front door open).

You have only one coin, so you can make only one call.

You have two neighboring houses.

Two persons live in one of them and three persons live in the other.

Both their telephones have answering machines.

Which one of the two numbers will you call?

The risk is that nobody will be at the home you call and the coin will be lost when the answering machine starts.

You could make an aleatory choice, but you could also make the following decision: "*Because the chances for somebody to be home are bigger in the case of house with three persons, I will call there.*"

Thus, you have made a decision based on your own comparison of probabilities.

Of course, the only information taken into account was the number of persons living in each house.

If other additional information—such the daily schedules of your neighbors—is factored in, the probability result might be different and, implicitly, you might make a different decision.

In the previous example, the estimation can be made by anyone because it is a matter of simple counting and comparison.

9

But in most situations, a minimum knowledge of combinatorics and the calculus of probabilities are required for a correct estimation or comparison.

Millions of people take a chance in the lottery, but probably about 10 percent of them know what the winning probabilities really are.

Let us take, for example, the *6 from 49* system (six numbers are drawn from 49 and one simple variant to play has six numbers; you win with a variant having a minimum of four winning numbers).

The probability of five numbers from your variant being drawn is about 1/53992, and the probability of all six numbers being drawn (the big hit!) is 1/13983816.

For someone having no idea of combinations, these figures are quite unbelievable because that person initially faces the small numbers 5, 6 and 49, and does not see how those huge numbers are obtained.

In fact, this is the psychological element that a lottery company depends on for its system to work.

If a player knew those figures in advance, would he or she still play? Would he or she play less often or with fewer variants? Or would he or she play more variants in order to better the chances?

Whatever the answers to these questions may be, those probabilities will influence the player's decision.

There are situations where a probability-based decision must be made—if wanted—in a relatively short time; these situations do not allow for thorough calculus even for a person with a mathematical background.

Assume you are playing a classical poker game with a 52-card deck.

The cards have been dealt and you hold four suited cards (four cards with same symbol), but also a pair (two cards with same value).

For example, you hold 3♣ 5♣ 8♣ Q♣ Q♦. You must now discard and you ask yourself which combination of cards it is better to keep and which to replace.

To achieve a valuable formation, you will probably choose from the following two variants:

– Keep the four suited cards and replace one card (so that you have a flush); or

– Keep the pair and replace three cards (so that you have "three of a kind or better").

In this gaming situation, many players intuit that, by keeping the pair (which is a high pair in the current example), the chances for a Q (queen) to be drawn or even for all three replaced cards to have same value, are bigger than the chance for one single drawn card to be ♣ (clubs).

And so, they choose the "safety" and play for "three of a kind or better".

Other players may choose to play the flush, owing to the psychological impact of those four suited cards they hold.

In fact, the probability of getting a flush is about 1/5.2 (19 percent) and the probability of getting three of a kind or better is about 1/15.7 (6.3 percent), which is three times lower.

In case you are aware of these figures beforehand, they may influence your decision and you may choose a specific gaming variant which you consider to have a better chance of winning.

But if you are in a similar gaming situation (you hold four suited cards and a pair) in a 24-card deck draw poker game, the order of those probabilities is reversed: the probability of getting a flush is 1/9.5 (10.5 percent) and the probability of getting three of a kind or better is 1/2.2, which is almost 50 percent and may help you to determine to keep the pair.

This is a typical example of a decision based on probabilities in a relatively short time.

It is obvious that, even assuming you have probability calculus skills, it is impossible to calculate all those figures in the middle of the game.

But you may use results memorized in anticipation obtained through your own calculations or picked from tables of guides containing collections of applied probabilities.

In games of chance, most players make probability-based decisions as part of their strategy, especially regular players.

The chapter titled *The Probability-based Strategy* provides a precise definition of this kind of strategy and a mathematical demonstration of the fact that, at least at a theoretical level, such a strategy is optimal.

The examples shown thus far were not chosen randomly.

They demonstrate the huge psychological component of our interaction with probabilities, especially in making decisions.

In addition to this practical aspect of interaction with chance and percentages, both laymen and scientists are fascinated by probability theory because it has multiple models in nature.

It is a calculus tool for other sciences and the probability concept has major philosophical implications as well.

Returning to the immediate practical aspect, whether we have mathematical background or not, whether we know the precise definition of the concept or not, or whether we have calculus skills or not, many times we make decisions based on probabilities as a criterion, sometimes without realizing it.

But this criterion is not obligatory. We may use it as a function of intuition, as a personal principle of life, or among other subjective elements.

Those making no decisions based on figures are those who run the risk unconditioned by a certain threshold and, several times, the result is favorable for them.

But statistics show that probabilities are most often taken into account, whether in a simplistic or professional manner.

All these are well-founded reasons for creating a probability guide addressed to average people who have no solid mathematical background.

This book succeeds in being a teaching guide addressed to beginners who want to acquire probability calculus skills, but is also a large collection of precalculated results that anyone can use directly, especially in gambling.

Although the main goal of this guide is practical, we also insisted on providing a rigorous review of probability concepts, whose lack makes the beginner subject to false interpretations and application errors.

The teaching material has been thus structured for people without a mathematical background to be able to picture a sufficiently clear view on the notions used and to solve the applications through an accurate framing of the problem and a correct use of calculus algorithms, even without reading the pure mathematics chapter titled *Probability Theory Basics*.

Those who feel that the high level mathematics is above their comprehension may skip that chapter without major repercussions on the practical goal, namely the correct application of calculus methods and algorithms.

This is possible because the presentation uses consistent references to mathematical notions through examples and natural models.

Moreover, the applications exclusively belong to finite probability fields, where the calculus algorithms and the reduced number of formulas used can be retained and applied without a complete and profound study of probability theory notions.

As a matter of fact, the reader needs only an elementary knowledge of mathematics from primary school to calculate odds and probabilities—operations with integer numbers and fractions (addition, subtraction, multiplication and division), the order of operations and elementary algebraic calculus.

Comfort with set theory and operations with sets is helpful, but all these notions can be found at the beginning of the chapter titled *Probability Theory Basics.*

Also, some knowledge of combinatorics notions and formulas is a great advantage.

Most of gambling applications use combinations and many times the probability calculus reverts to counting them.

But the lack of such knowledge is successfully compensated for because the guide contains a solid chapter dedicated to this mathematical domain and the information provided there can be understood by anyone.

The structure and content of the next chapters follows:

What Is Probability?

Reading this chapter is essential in accommodating with the probability concept.

Starting from varying the mathematical definition, the probability is structurally shown as a measure and also as a limit.

All mathematical notions referred to, as well as the main probability theory theorems, are exposed through examples and natural models.

We also talk about the philosophical interpretations and implications of the concept and about the psychological impact of interaction of humans with probabilities.

Probability Theory Basics

This is the strictly mathematical chapter that contains all the rigorous definitions that establish the basis for the probability concept, starting from set operations, sequences of real numbers, convergence, Boole algebras, and measures, to field of events, probability, conditional probability and discrete random variables.

The chapter contains only the main theoretical results, which are presented as enunciations without demonstrations, but also contains many examples.

As we said before, reading this chapter is not obligatory to understanding the practical goal of calculus, but following it in parallels when running through the first chapter is useful for those having a minimal mathematical background.

Combinatorics

Combinatorial analysis is an important calculus tool in probability applications and this is the reason why it has a dedicated chapter.

This chapter contains the definitions of permutations, combinations and arrangements, along with their calculus formulas and main properties.

A lot of examples and solved and unsolved applications complete the theoretical part.

Beginner's Calculus Guide

This part of the work is in fact the main teaching material required by the beginner who wants to correctly apply probability calculus in practical situations.

The exposure of methods and applications is mostly algorithmic to enable the reader to easily follow the steps to be executed and to avoid the application errors.

Framing the problem, establishing the information to be taken into account, the correct enunciation of the events to measure, the adequate calculus method, the formulas to use, the calculus, all are explained and exemplified in depth and in an accessible manner.

Solved applications and suggestive examples are shown throughout this chapter.

You will also find here the most frequent errors in application or calculus pointed out to help beginners avoid them.

As in other chapters, the examples and solved applications mostly belong to games of chance, where the knowledge actually acquired has immediate application.

The probability calculus is explained and applied for the finite case (finite field of events), where the practical situations suitable for application are more numerous (in gambling and everyday life) and probability has a higher accuracy as a decisional criterion.

Probability Calculus Applications

This chapter verifies the theoretical and application knowledge acquired in the previous chapters.

It contains a collection of solved and proposed elementary problems that usefully and even necessarily complete the previous theoretical chapters.

Applications and Results for Games of Chance (Slots, Roulette, Blackjack, Texas Hold'em Poker)

Games of chance stand for the most relevant experiments that generate probability measurable events, and counting these probabilities is an important part of players' strategy.

This chapter has a double goal: continuing the series of applications of acquired knowledge and presenting a large collection of numerical results that players can use in frequent gaming stages and situations.

Each section corresponds to a game of chance and starts with a description of the respective game and its rules, and then goes on with the general framing of probability problems within the respective game.

Next you will find some completely solved applications and a collection of precalculated results that are structured on categories of gaming situations specific to respective game.

All numerical results are illustrated in tables and listings at the end of each section, from where they can be extracted and directly used by interested players.

The Probability-based Strategy

The preponderant usage of probability as a decision-making criterion naturally generates the need for theoretical motivation.

This chapter is an essay containing the definition of probability-based strategy and mathematical demonstrations showing that such strategy is theoretically optimal.

Various Applications

This chapter continues the series of useful applications for probability calculus training, with proposed problems whose difficulty level grows gradually.

The applications come from games of chance and also from everyday life.

WHAT IS PROBABILITY?

The goal of this chapter is to provide the reader who has no mathematical background with a sufficiently clear image of the probability concept, and how, in its absence, the approach of proper calculus becomes predisposed to errors, especially as they relate to the correct initial framing of problems.

This image should catch the basic structural aspects as well as properties that become manifest when we ascribe events to statistics and probability in daily life.

We will also talk about the psychological and philosophical implications of this concept, trying to delimit the mathematical term from the content of the common word *probability*.

Obviously, any explanation and presentation of the probability concept, even for people without a mathematical background, cannot overlook the mathematical definition. Thus, we will start with this definition and try to rebuild it step by step from its constituent parts. This rebuilding will be on the basis of particular cases and suggestive examples.

Such an attempt is a must because of the huge psychological impact of statistics and probabilities in people's daily lives.

Due to a natural need that is more or less rigorously justified, humans consistently refer to statistics; therefore, probability has become a real decision-making tool.

In this chapter we discuss in detail this major psychological component of the probability concept.

The fact that people estimate, calculate and compare probabilities with the purpose of making decisions, without knowing exactly the concept's definition or without mastering the proper calculus, automatically generates the risk of qualitative interpretation errors and the errors that come of using the figures as decision-making criterion as well.

Psychologically speaking, the tendency of novices to grant the word *probability* a certain importance is generally excessive in two ways: the word is granted too much importance—the figures come to represent the subjective absolute degree of trust in an event occurring—or too little importance—so many times, an equal sign is put between *probable* and *possible* and the information provided by numbers is not taken into account.

We use here the term *word* instead of *concept* deliberately when talking about probability.

To have a good understanding of what *probability* means and implies, we do not refer to the mathematical definition exclusively, but also to how this notion is perceived in a nonacademic environment for people having low or average knowledge level.

This approach is necessary because probability, as both a notion and a tool, has huge psychological implications when interacting with human concerns in daily life.

Words and concepts

For a correct interpretation of all information we collect to study the various objects of knowledge, is absolutely necessary to be able to make the distinction, so many times hardly discernible, between word and concept.

A **word** is a graphic and phonetic representation of a category of objects that are subjects of judgments or of human communication.

Thus, a word identifies a group of objects from the surrounding reality, whether physical, perceived or abstract.

For example, we can assign to the word *apparatus* the set of objects (television, radio, microphone, computer, the group of internal human organs that work the digestion, and so on). (This is in fact a set of words, which, by an abuse of denotation, has been presented as set of physical objects, namely, the set of all such kinds of objects existing). The denotation *apparatus* represents that entire set.

As another example, we can assign to the word *ball* the set of objects {the round object used in soccer, the round mobile object in pool (snooker) game, a celestial body, and so on}, whose representative is the word itself.

The word is practically a symbol, a denotation that is assigned to the set of objects it represents.

Language arose and evolved during history along with the need for human communication, but the status of a word always remains the same: a symbol indispensable to communication but deprived of conceptual content.

The word is not an object itself, but a symbolic representation of other objects from the surrounding reality, having the exclusive purpose of communication.

Words are used in common nonacademic language as communication symbols. The transmission of a certain word automatically refers to the entire set of objects that word represents.

These sets of objects are generally accepted by the community using a specific language, in the sense of majority.

This agreement is neither written nor officially stated by a certain organization invested in this matter, nor is it suggested by any academic society. While most people refer to dictionaries to gain the definition of a word to avoid argument about its usage, in this case, it is the practical result of free human communication throughout history.

A **definition** is a grammatical sentence delimiting or extending a set of objects that can be attached to a word by enunciating the properties specific to the objects it describes.

A definition can be attached to a word (or group of words) when the set of objects it represents is not unanimously accepted or there are doubts about this matter within a community or even between two interlocutors.

On the other hand, a definition can be attached to a group of objects in order to simplify the ulterior communications that refer to those objects.

The sentence the definition consists of enunciates the properties specific to the objects involved.

At a scientific level, a definition stands for the base of any theory because a theory cannot operate with undefined objects.

The objects studied must be well defined initially before they can become the subjects of judgments and communications.

At a common nonacademic communication level, the need for a definition appears many times, especially when there is a risk of confusion about the objects that are subject of communication.

Generally, the elaboration of a definition comes after an interlocutor's request.

The grammatical predicate of the sentence a definition consists of can be *is called, is named, is defined* or *is*.

Therefore, the definition assigns to a word (or group of words) a sentence delimiting or extending the associated set of objects or assigns a word (or a group of words) to a set of objects having specific properties.

Examples:

1) Definition: A *ring* is a circular shaped object made of any material.

This definition attached to the word *ring* extends a possible set of objects previously assigned, such as the set of circular shaped jewels worn on the finger.

Still, this definition delimits a possible set of objects previously assigned, such as a set of all natural or artificial circular shaped objects.

2) Definition: A *square* is a rectangle having diagonals of the same length.

This mathematical definition assigns the word *square* to the objects, *rectangles whose diagonals have the same length,* that is, to any rectangles having this specific property.

The purpose for the elaboration of a definition is to illustrate an object or category of objects that are subjects of communication, upon which a theory is given out or which completes an existing theory, and to simplify communication by reducing the words required to communicate.

Definitions do not create new objects but draw certain objects from a larger category. They can eventually create new words (groups of words).

In a more abstract sense, we may say the word also represents a definition itself, if we take into account the set of objects it has attached.

Generally, however, this set cannot be covered entirely by enunciations standing for a compact definition.

As mentioned earlier, a definition is a sentence having a grammatical sense. We can further attach definitions to the various words that make up the sentence, and so on.

Any word can be defined in many subjective ways and its definition, in turn, also consists of words.

Thus, attaching a definition to a word does not confer on it a status that is higher than a simple symbol, as long as this definition contains only words.

But this becomes possible in scientific context, although in a limited way, due to philosophical matters.

Because a theory from an exact science operates only with well-defined objects, in that sense that all words contained in a definition are rigorously defined in turn, the words representing these objects almost always identify themselves with the associated objects, at least with respect to the elimination of confusion.

In these cases, the triple *word–attached definition–defined object* is called a *concept* or *notion*.

A **concept** is an abstract object whose definition contains only words that are well defined in the scientific sense, with direct reference to the specific science it belongs to.

The elaboration of a rigorous definition of the concept itself cannot be performed structurally by using other objects. The object referred to (the concept) can be only defined through itself:

A concept is an object whose definition contains only concepts (and not words having no scientific definitions attached).

This definition confers on *concept* an abstract characteristic from the beginning.

The concept as a definition also has a relative characteristic, because the quality of well defined makes sense only when related to the science, field or theory we operate within.

As a basic example, all mathematical concepts are abstract objects that are well defined through other rigorous mathematical notions.

Example:

Definition: A set *A* is called a *countable set* if there exists a bijective function from *N* (the set of natural numbers) to *A*.

Within mathematical analysis, to the words *countable set* a definition is attached and this definition contains only words that are well defined in mathematics, such as *function* and *bijective*.

Thus, the group of words *countable set* acquired the status of a mathematical concept (the words *countable set* + their definition + the objects they represent).

We may identify the words *countable set* with the mathematical concept defined above, but only within a communication in a scientific environment.

In any nonacademic communication, these words never improve their symbol status because arbitrary interpretation and even confusion are possible.

Let us take again a previous example, to which we assigned the word *ball a* set of objects (the round object the soccer sport uses, the round mobile object of pool (snooker) game, a celestial body, and so on).

In mathematics (geometry and topology), the ball represents a well-defined concept:

Definition: In R^3 space, if $C \in R^3$ and $r > 0$ is a real number, then the set $\{x \in R^3, d(x,C) \le r\}$ is called the closed ball of centre C and radius r.

Any set with this form is called a closed ball.
($d(x,y)$ is the distance between points x and y; the similar definition of the open ball is obtained by replacing $\le$ with $<$; a set that is a closed ball or open ball is called *ball*).

Thus, the word *ball* used in common language has acquired a concept status as a result of a rigorous mathematical definition being attached.

Obviously, the mathematical concept of ball also belongs to the set of objects the word ball represents (the round object the soccer sport uses, the round mobile object of pool (snooker) game, celestial body, the mathematical concept of ball, and so on).

This membership is valid for any word that becomes a concept: the representative set of objects attached to respective word automatically contains the abstract object with the same name.

We will not insist further on a philosophy of definition and language matters.

This brief presentation of terminology and general definitions has as its main goal to catch the structural differences between *word* and *concept*, which are necessary to clarify the mode of perception and interpretation of mathematical notions by the people without a scientific background.

Taking into account everything so far presented, we must bear in mind the following:

– A word is a simple symbol that is necessary for communication that represents a set of objects from the surrounding reality, either physical or abstract;

– Definitions can be attached to words; these definitions may be more or less rigorous and more or less subjective;

– Definitions are sentences also consisting of words that illustrate one or more objects with specific properties;

– Concept is one abstract object, whose definition contains other previously defined concepts; the definition attached to a concept contains only words that are well defined in within the scientific field it belongs to;

– In the scientific context, the concept identifies itself with the word it represents, and at that communication level;

– The concept as an object is part of the set of possible objects the respective word represents.

Let us return to the example of the mathematical definition of ball.

A ball in the three-dimensional space R^3 has been defined by using the classical orthogonal distance (metric): the distance between two points x and y with coordinates (x_1, x_2, x_3), respectively (y_1, y_2, y_3) is

$$d(x,y) = \sqrt{(x_1 - y_1)^2 + (x_2 - y_2)^2 + (x_3 - y_3)^2} \, .$$

But there are generalizations of this notion. The immediate generalizations are obtained by extending the space R^3 to an arbitrary metric space, structured with an arbitrary metric.

A metric space is a couple (X, d), where X is a set and d is a metric on X. The definition of metric is:

A function $d : X \times X \to R_+$ is called metric on X if it fits the axioms:

   a) $d(x,y) = 0 \Leftrightarrow x = y$

   b) $d(x,y) = d(y,x), \forall x, y \in X$

   c) $d(x,y) \leq d(x,z) + d(z,y), \forall x, y, z \in X$.

We can now enunciate the generalized definition by calling the newly defined object *ball*, too:

Definition: Let (X, d) be a metric space, $x \in X$ and $r \in R_+$. If we denote $B(x,r) = \{y \in X, d(x,y) < r\}$ and $B[x,r] = \{y \in X, d(x,y) \leq r\}$, then $B(x,r)$ and $B[x,r]$ are called the *open ball* of centre x and radius r, respectively, the *closed ball* of centre x and radius r.

Any subset of X having one of these two forms is called *ball*.

Thus, we have put in evidence two abstract objects, with the help of two definitions: one attached to the concept of ball in R^3 space and one attached to the concept of ball in a metric space.

Although one is the generalization of another, the two concepts are distinct, but they are represented by the same word, *ball*.

Although choosing the same word to represent the generalized concept has a solid theoretical motive in mathematics, this automatically generates the danger of confusion in communication and of the elaboration of incorrect judgments about objects merging because they are represented by the same word.

Referring to our concrete example, in an academic environment this danger is minimal, practically nonexistent, because any mathematician who communicates or develops a theory containing those two concepts will professionally relate them to the base-space respective definitions refer to.

But in nonacademic communication, the risk of miscommunication or error is much higher.

People without a mathematical background will operate with the word *ball* representing a set of objects from which the two concepts, *ball in* R^3 and *ball in a metric space*, may be a part or not, depending on subjective choice or the knowledge level of the person using the word ball.

We have then the same word *ball* with three different meanings: the word-symbol *ball* (with its attached set of objects), the concept *ball in* R^3 and the concept *ball in a metric space.*

At the common nonacademic communication level, the danger of incorrect transmission about the content and essence of circulated words naturally appears.

This can generate conflicts at the level of simple discussion between interlocutors, and at theoretic level as well, through the development of judgments or theories operating with inconsistent terms or even having elementary logical vices.

The conflicts generated by words that do not carry a uniquely defined or sufficiently solid content may appear both in daily communication and at the scientific level.

There are the so-called exact sciences in which the theories begin by defining some notions inconsistently and assigning to them words representing a much more extended content.

Moreover, by granting these notions properties belonging to other objects represented by the same word, conflicts appear as logical vicious circles or even paradoxes, as do scientific results that are apparently acceptable but are in fact erroneous.

The science considered to be the most exact, mathematics, contains unsolved paradoxes generated by the inconsistency in defining some concepts.

The most classical example at hand is the set notion. This notion stands for the base of any construction of a mathematical theory and is contained in almost every definition.

All mathematical concepts are defined through other concepts, which are rigorously defined at their turn, downwards to the concept of *set*. This is a so-called *zero degree* term that cannot be rigorously defined with the help of other concepts.

The set is described as a group or collection of distinct objects called elements of the set.

The set and any element of it are distinct objects from different mathematical categories.

Although unanimously accepted, this still is not a conceptual definition. The words *group, collection* and *objects* do not have the status of a rigorously defined concept.

For example, the word *objects* without status as a concept allows us to choose any representative from the attached set, particularly *the sets*.

We may then consider *the set of all sets* as a chosen set. But, after thorough observation, we see that this set itself is also one of its elements!

This paradox classifies from the beginning the definition of set as being nonrigorous, even it does not generate logical conflicts in any theories built upon it.

These conflicts generated by language cannot be avoided as long as science finds no form of communication superior to one using only words.

Such communication would transmit pure information directly to the human brain's processing systems (with the condition that receivers suitable for receiving such information exist) by skipping the intermediate step of representation through symbols (words).

Proving or disproving the existence of such an ideal communication system remains a study problem to which disciplines like the philosophy of science, metaphysics, mathematics, physics, parapsychology or even religion may contribute.

Until then, the word remains the only accessible and viable communication mode, even though it is an imperfect tool with respect to the interpretation of the intrinsic content of certain abstract objects that interact with us at the cognition level.

Mathematical models

To study the various objects of physical reality and their properties, the exact sciences develop theories based on models of these objects.

A model is an ensemble of scientific concepts reproducing the studied object or objects at a theoretical level.

Creating a model and developing a theory upon it generate an equivalent principle with the reproduced object (objects).

Physics and mathematics especially operate with models to study the surrounding reality by discovering as many properties of the constituent objects as possible.

The need for elaboration of models is a natural one from a scientific point of view.

This need is generated by the fact that experience is not sufficient to know the studied objects and their evolution (structure, properties, behavior, and predictions), even if only with regard to perception.

Thus, theory creates models that reproduce the studied objects or phenomena from the perspective of the essential aspects that a respective theory uses to find new properties.

As a practical matter, the physical object is reduced to a theoretical concept on whose base an ulterior theory can be built.

Results obtained theoretically are applicable to the modeled physical object, which makes possible the acquisition of additional information about the object as well as the making scientific predictions.

As an example, the entire field of celestial mechanics was discovered by using physical and mathematical models that reproduced the moving celestial bodies and their mutual interactions.

Because these bodies are intangible and eventually visually perceptible, their laws of motion cannot be found through experience or through long observation.

A model created by physics was needed, through which the celestial bodies were reduced to simple material mass points interacting within gravitation principles and laws.

Thus, a physical model was created that is equivalent to the ensemble of real objects with respect to the motion caused by the forces involved.

Mathematics then established motion equations and trajectories, based on the applied physical laws, thus creating in turn a model of the moving objects.

Based on these equations, we can reproduce and predict the trajectory of any celestial body from a given ensemble as a function of the initial coordinates and other parameters such its mass and the mass of other objects gravitationally interacting with it.

Observation confirmed the results obtained theoretically through the study of each respective model, which was also valid for the corresponding physical phenomenon.

The equivalence established between the real object and the attached theoretical model is not total.

Elaboration of a model assumes approximations and reductions of the physical object by illustrating only a few of the properties the theory considers necessary to work on.

For example, to calculate (approximate) the volume of Earth, a scientist chooses the mathematical model of a sphere, whose properties are used to work out the proposed calculus.

The object Earth is thus reduced through approximation to the mathematical object (concept) of sphere.

Properties of the physical object, such as surface and shape irregularities, are not taken into account because they are considered unessential to the mathematical calculus to be worked out.

Another approximation may use the geometric model of a spheroid, by ignoring those same physical properties because they are considered negligible.

If we propose not only to approximate the volume of Earth, but also to make a plot of its trajectory around the Sun, then the object Earth would be reduced as a model to a simple material mass point, as would the Sun, in order to apply the gravitational laws for finding the motion equation. The geometric model of a sphere or spheroid becomes no longer helpful in this attempt.

The creation of scientific models makes the transfer of information from experience to theory at the cognitive level.

The deduction process (achieving theoretical results for the model in use) becomes an inductive one when the theoretical results are transferred and assigned to the modeled object.

Thus, the information is transferred backwards, from theory to experience, after it is deductively processed at the theoretical level.

This double assignment process creates an equivalence between the object or phenomenon studied and the model created by science, but only with respect to the properties studied.

The real object is different from the scientific model attached and contains much more information than the last one.

Therefore, the real object has additional properties that are not theoretically found in the model used.

As an example, to study the motion of a planet couple; that is, a planet and its natural satellite, progressively approaching one another, and to predict the time remaining until the bodies merge, a physical-mathematical model is attached, namely, two mass spheres in gravitational interaction.

By using this model and data like the distance between the two spheres, their radiuses and masses and their velocity at a certain moment, a prediction for the time remaining until they merge is possible through the use of a rigorous mathematical calculus.

The use of the properties of two mass spheres in gravitational interaction is enough to make that prediction.

But the two celestial bodies also have other physical properties that are not visible on the created model, such as composition, density, relief, and the like.

These are not taken into account because the applied theory had no need for them.

Here is another example, known in mathematics as a recreational paradox, which reveals that the equivalence between a model and the phenomenon reproduced is not total and may result in erroneous results:

A man must walk the distance between points A and B. To reach point B, the man must first walk half the distance AB, then must walk half the distance remaining, then again half the distance remaining, and so on, to infinity. It can be concluded that the man will never reach the final point B.

Obviously, experience has proved the conclusion wrong. But where is the error?

Well, we do not deal here with a vicious logical judgment and there is no well-hidden error in the context. The real error has a conceptual dimension; namely, it is an error in the transfer of information from the theoretical model to the physic reality.

The mathematical model of the described phenomenon, on which the judgment was performed, follows:

We have a segment [AB] with L denoting its length.

Consider the midpoint of this segment, which is a geometric point. Denote it by A_1.

The distance from A_1 to B is L/2. Consider now the midpoint of segment $[A_1 B]$ and denote it by A_2.

The distance from A_2 to B is L/4.

By continuing to halve the segment left and similarly denoting the midpoints, we obtain a sequence of points $(A_n)_{n \geq 1}$.

It can be mathematically proved that this sequence converges toward point B. But point B is still not reached (a natural number n

such that $A_n = B$ does not exist), because the distance from A_n to B is $L/2^n$, which is not null, for every n.

This is the judgment made on the mathematical model, which is entirely correct.

The error comes from the fact that the chosen model does not represent the problem well enough. The motion is discretely represented, while in fact it is a continuous one in reality.

The sequence $(A_n)_{n \geq 1}$ does not represent the motion, but is a simple subsequence from the continuous set of covered points, which also include terminal point B.

This is, in fact, a reflection of the difference between discreteness and continuum in mathematics.

Moreover, the real phenomenon and the chosen mathematical model also have an essential difference: while the *cuttings* halving the distance are made through geometric points within the model, in the physical case these are made through human steps.

The step has a certain length corresponding to the sole, which covers the distance remaining to point B at a certain moment of the walk.

Within the mathematical model, the equivalent of the step, namely the point, has a null length. This is exactly why point B will never be reached by the sequence of points mentioned, while in reality, physical steps will reach the terminal point after the last halving that leaves a distance lower than the sole's length.

Thus, even if the problem was formulated for the chosen model to correctly describe the motion (the distance to be covered through isolated points and not continuously), this model is still not sufficiently illustrative because the length of the step should be zero, which is impossible.

This is an eloquent example in which the transfer of the theoretically results obtained from the model to the physical phenomenon is invalidated by reality.

This is because the model created is not totally equivalent to phenomenon described, which contains additional information.

The prediction is one of the main purposes of sciences and through models the theory succeeds in obtaining scientific predictions on the real objects and their evolution.

As we saw in the example, when a physical object is reduced to a theoretical model, although the construction of theories developed on it are not logically altered in any way, the loss of information belonging to studied object, as well as the definitions of concepts involved in the model, confer to the cognition through models a strong relative characteristic.

In mathematics, although that relative characteristic is maintained at the stage of transfer of results from the model to physical objects or phenomena, there is an accuracy of transfer that is higher than in other sciences.

Usually, mathematics does not make this transfer directly; it is a tool used by other sciences, especially physics, which creates scientific models for the surrounding reality.

Mathematics can only be contested at the philosophical level, where all become relative.

Finding the mathematical properties of studied physical objects, as well as calculus problems of any kind are completely solvable through a model attachment, even that relativity generated by the transfer of theoretical results from the model to real object is maintained.

But this relativity converges toward the knowledge limits and is exclusively a matter of philosophy of thinking.

Mathematics offers the lowest conceptual and computational error range, in condition that these limits are acknowledged and accepted.

We made this brief incursion into the basic problems of scientific representation through models, as well in the philosophy of definition, to familiarize the reader with the idea that processing information at the cognitive level is not exclusively focused on the structure and content of information, but also to go deeply into the source of each constituent element and even to the mode of human perception of all these.

For any problem of scientific study, the conceptual group *word–notion–model* must be taken into account as a whole unit and individually as well in order to make the necessary distinctions to avoid the conceptual errors in judgment processes and even in simple communication.

Returning to *probability*, its study problems will entirely submit to all the examples presented previously.

As already mentioned, the explanation of the probability notion cannot leave aside the rigorous mathematical definition on which we will focus further.

Probability will be presented both as a simple word and as a notion, along with the mode of perception of both for people having no scientific background because, unlike other mathematical concepts, this one has a huge psychological impact in people's daily lives.

Probability – the word

In the current language of a nonacademic environment, the words *probability* and *probable* are frequently used by a large number of people, and those people assign to these words meanings that partially or totally refer to the mathematical definition of the respective concept.

Practically all subjectively attached definitions catch the measurable, quantitative aspect of the notion, which can only be described at the mathematical level, and even this means only the simple attachment of some numbers.

The common definition of probability, accepted by the large majority at a nonacademic communication level, is the trivial mathematical definition of probability on a finite field with equally possible elementary events.

Of course, there are two categories of people—those who know this definition (with or without a mathematical background) and those who interpret it unrigorously and subjectively.

But for both categories, the following elements are generally known and accepted:

– Probability is a positive number less than or equal to 1, which can be assigned to the possibility of an event occurring as the result of an experiment.

– This number is the result of a numerical calculus or just a simple observation and is usually expressed through percentages.

Being less than or equal to 1, it comes from a proper fraction.

– Probability 1 is assigned to a sure event.

Obviously, these elements do not entirely compose the mathematical definition of probability, nor even the elementary one.

Moreover, the terms *experiment* and *event* are used as simple dictionary words without a conceptual status.

People with a minimal mathematical background will assign to the word the classical definition of probability on a finite field with equally possible elementary events:

Definition: Let A be an event. Call the probability of event A the ratio between the number of cases (tests) that are favorable for A to occur and the number of all equally possible cases ($P = m/n$).

At this moment, we limit ourselves to exemplifying this definition through few simple experiments, without describing the constituent terms.

We will do this in the paragraph dedicated to the reconstitution of probability concept.

1) In the die rolling experiment, the elementary events as sets are $\{1\}, \{2\}, \{3\}, \{4\}, \{5\}, \{6\}$.

The probability of such an event (any of them) is 1/6.

If we look for the probability of the event *the die shows 2 or 3,* represented by set $\{2, 3\}$, let us observe that it is a compound event and can occur through two tests (cases) generating events $\{2\}$ and $\{3\}$.

The probability of event $\{2, 3\}$ is the ratio between the number of these favorable cases (two) and the number of all equally possible cases (six).

2) An urn contains three white balls and four black balls. Someone randomly extracts a ball from the urn.

The probability of extracting a white ball is 3/7 (three tests are favorable for extracting a white ball from a total of $3 + 4 = 7$ equally possible tests).

3) From a 52-card deck, someone randomly picks a card.

What is the probability for this card to have a value higher than 10?

The number of cases that are favorable for event *the card has a value higher than 10* is given by the number of cards having this property. These are: J♠, J♣, J♥, J♦, Q♠, Q♣, Q♥, Q♦, K♠, K♣, K♥, K♦, A♠, A♣, A♥, A♦, a total of 16.

Picking cards represents equally possible elementary events, so the probability of that event is 16/52, this is 4/13, or about 30.76 percent.

This elementary definition, or grammatical variants of it, is generally attached to the word *probability* by people having a minimal mathematical background.

Operating with the word *probability* given this definition is correct as long as does not extend through language to events belonging to more complex probability fields where, mathematically and conceptually, it is no longer valid.

Here is where the conflict between word *probability* and the concept having same name occurs. The mathematical concept is much more complex, so we will reconstitute and explain its definition later in this chapter.

Also, performing calculations based on the classical definition in a context where this cannot be applied due to mathematical reasons leads to erroneous numerical results.

Regarding the word *probable*, in the current language this word represents a quality to assign to a certain event related to its probability.

Terms such as *probable, less probable,* and *very probable,* when used to describe events, become subjective categorizations having numerical probability as a criterion.

Being numerical, the ordering criterion is automatically a subjective one.

Any definition of these words can only be built quantitatively and therefore subjectively.

We call an event probable if the value of its probability (observed, communicated or calculated) exceeds a certain subjectively chosen threshold (among them, 50 percent is a frequently used threshold).

Many times, the quantitative criterion of assigning the quality of probable is not a matter of direct estimation of probability, but of observation of practical statistics.

An event is said to be probable when it occurred under similar conditions, according to statistics, in a certain number of cases previously registered by statistics until a respective moment.

This statistical criterion also assumes the evaluation of a ratio, which is not probability any more, but relative frequency.

Obviously, the subjective choice of a threshold here is part of the criterion, too.

In mathematics, the notion of *probable* does not exist, so the use of this word is reduced to a simple common communication of properties subjectively attached to events.

As we may observe, the calculus formula from the classical definition of probability is very easy to apply. Its algorithmic application consists of:

1) Textually defining the event whose probability we are searching for;

2) Counting all possible tests having as a result the occurrence of respective event (the favorable cases) (m);

3) Counting all equally possible tests (n).

4) Doing the ratio of numbers obtained at steps 2) and 3) ($P = m/n$).

Let us consider another simple example:

From a 52-card deck, a person randomly picks one card. Let us calculate the probability for this picked card to have the clubs symbol (♣).

We apply the classical definition of probability, because we can consider the aleatory extraction of cards from the deck as equally possible events.

By following the calculus algorithm, we have:

1) The event to study is *picking a card with the clubs symbol.*

2) There exist thirteen clubs in the deck (one for each value A, 2, 3, 4, ..., etc.). The extraction of clubs can occur through thirteen possible tests (extraction of each of the thirteen clubs cards) (m = 13).

3) The total number of equally possible tests is 52 (extraction of each of the 52 cards) (n = 52).

4) The ratio is $P = m/n = 13/52 = 1/4 = 25\%$.

In this way, to an event we can attach a number less than or equal to 1, which we call the probability of that event (in an exclusively numerical sense, only on condition that all elementary events are equally possible).

By the classical definition of probability, attaching this number to an event means the construction of a simple function on the set of events, without additional significations.

The complex interpretation of this function and the knowledge of its properties will become possible when we explain the complete definition of probability.

Returning to the card deck example, let us change the hypothesis a little bit, by saying that, after a respective card was picked, we accidentally saw one card from the deck, which was 5♦ for example.

The problem remains the same—to calculate the probability for that picked card to be of the club suit.

Obviously, the event whose probability we calculate is expressed through the same word, namely, *picking a card with the clubs symbol.*

In addition, we have the information provided by viewing one of the cards left in the deck. Let us see how this information changes the numerical results of the calculus algorithm based on the classical definition.

At step 2), the number of possible tests that are favorable for the occurrence of the studied event remains thirteen (each of the thirteen clubs cards could be the picked card) ($m = 13$).

At step 3), the number of equally possible test is 51 instead of 52 because the test of picking the card 5♦ is impossible (the card was seen in the deck after extraction, so it cannot be the one extracted) ($m = 51$).

Therefore, the ratio calculated at step 4) is $P = 13/51$, which is bigger than $13/52$.

How is it possible for the same event to obtain two different probabilities, as long as probability, according to the classical definition, is a function? (See the mathematical definition of function in the next chapter.)

The answer is very simple: although the two events (corresponding to the two problems) whose probability was calculated are textually described as the same, they are not identical.

To comprehend this, we should go deeper into the notion of *event*, which is used as a simple word thus far.

Event as a mathematical notion cannot be isolated from a certain context of its definition.

When we extend the definition of probability as a function on a set of parts, we see that it only makes sense in a predefined field of events, which forms a probability field.

Many probability fields that are different through their field of events mean different probabilities (as functions).

This is an essential aspect that puts in evidence the mathematical relativity of probability concept.

In the two previous examples, the fields of events were different, as were, implicitly, their probability fields.

But to reach the extended mathematical definition of probability, we must further explain some notions this definition is built through, like Boole algebra, sequence of sets, tribe, measurable space, measure, and the like.

The concept of probability

Initially, probability theory was inspired by games of chance, especially in 17th century France, and was inaugurated by the Fermat–Pascal correspondence.

However, its complete axiomatization had to wait until Kolmogorov's *Foundations of the Theory of Probability* in 1933.

Over time, probability theory found several models in nature and became a branch of Mathematics with a growing number of applications.

In Physics, probability theory became an important calculus tool at the same time as Thermodynamics and, later, Quantum Physics.

It has been ascertained that determinist phenomena have a very small part in surrounding nature.

The vast majority of phenomena from nature and society are stochastic (random).

Their study cannot be deterministic, and that is why hazard science was raised as a necessity.

There are almost no scientific fields in which probability theory is not applied.

Also, sociology uses the calculus of probabilities as a principal tool.

Moreover, some commercial domains are based on probabilities (insurance, bets, and casinos, among others).

Probability as a limit

We initially present the probability concept as a limit of a sequence of real numbers.

Although this is a particular result (a theorem, namely *The Law of Large Numbers*) and not a definition, it confers to probability a structure on whose base the comprehension of the concept becomes clearer and more accessible to average people.

Also, the perception of probability as a limit diminishes the risk of error of qualitative interpretation with regard to the real behavior of random phenomena, given the mathematical model.

Although in the strictly mathematical chapter the chronology of presentation of the notions and results is the natural one from a scientific point of view (definitions – axioms – theorems), here we deliberately reverse this order by explaining the probability concept in a particular case, as a limit, for a sequence of independent experiments that aim at the occurrence of a certain event.

This presentation mode is chosen with a didactic purpose because the notion of *limit* is easier to explain, visualize and assimilate at a nonacademic level than the complete definition of probability based on Kolmogorov's axioms.

This last one is explained for beginners' understanding in the next section.

In addition, in mathematics, the status of *definition* and *theorem* may commute within the same theory, without affecting the logic of the deduction process.

This means the deduced property of a mathematical object can stand for its definition itself, and vice versa.

Here is a simple example: In geometry, if we start from the definition of square being *a rectangle with all legs having the same length,* we can immediately prove that, in a square, the diagonals make 45-degree angles with the legs, this being a deduced result.

If we change the definition by calling square *a rectangle whose diagonals make 45-degree angles with the legs,* it can be easily proved that all legs of a square have the same length, which is the first definition of a square.

Thus, the definition of a square and the theorem that states a property of it are commuting, but both definitions determine the same mathematical object, called *square.*

Let us go back to the concept of the probability of an event.

To describe the probability as a limit, we explain and exemplify the elementary notions of sequence and limit of a convergent sequence of real numbers.

A sequence is an infinite enumeration of objects, some of which may be distinct or may repeat.

In mathematics, *enumeration* means a function defined on N (the set of natural numbers), with values (images) in a set of objects (the range of the function).

If the function's domain is N, then through this function, an element from the range is associated with each element of the domain (see the section titled *Functions* in the mathematical chapter).

The associated element from the range is called term of the sequence.

If A is an arbitrary set, a function from N to A associates:

to number 0 – an element $a_0 \in A$

to number 1 – an element $a_1 \in A$

to number 2 – an element $a_2 \in A$

....

to number n – an element $a_n \in A$

....

and so on toward infinity (∞).

The elements $a_0, a_1, a_2, ..., a_n, ...$ (in enumeration order) form a sequence, also denoted by $\left(a_n\right)_{n \geq 0}$ or $\left(a_n\right)_{n \in N}$ or $\left(a_n\right)_{n=0}^{\infty}$.

Index n is called the rank of term a_n and represents the term's order number within the sequence.

A sequence must not be confused with the set of its terms because they are different mathematical objects (function, respectively set).

A sequence assumes an order of its constituent terms and, in addition, the terms of a sequence can repeat, while the elements of a set cannot.

There exist or we can build sequences of numbers, of sets, of functions and sequences of however many complex objects.

Here are few examples of simple sequences:

1) $0, 2, 4, 6, 8, \ldots, 2n, \ldots$ is the sequence of natural even numbers, which can be denoted by $(2n)_{n \geq 0}$ (expression $2n$ is the one that generates all the sequence's terms and is called general term).

Observe that the terms of this sequence are continuously increasing (it is an increasing sequence) and they cannot all be included in a bounded interval (the sequence is unbounded), approaching infinity as their rank goes up.

We say the sequence has the limit ∞ or tends to infinity.

For a sequence to tend to infinity, it is not necessary for the sequence to be increasing.

As an example, the sequence $0, 1, 0, 2, 0, 3, 0, 4, 0, 5, \ldots$ is not increasing, but it still tends to ∞.

Besides, if a sequence tends to infinity, it is by definition unbounded.

For the exact definitions of a bounded, increasing or decreasing sequence, see the section titled *Sequences of real numbers. Limit* in the mathematical chapter.

2) $0, 1, 1, 2, 3, 5, 8, 13, 21, \ldots$ is an increasing sequence of natural numbers in which a term is obtained by summing the previous two terms (called a Fibonacci sequence). This sequence has the limit ∞, too.

3) $\dfrac{1}{2}, \dfrac{1}{3}, \dfrac{1}{4}, \dfrac{1}{5}, \ldots$ is a decreasing sequence of rational numbers (fractions).

4) $\{1\}, \{1, 2\}, \{2\}, \{1, 2\}, \{3\}, \{1, 2\}, \{4\}, \{1, 2\}, \{5\}, \ldots$ is a sequence of sets.

We may create any sequence by choosing its terms from the elements of a set of objects in the same category.

If set A (the range of function representing the sequence) is a subset of real numbers ($A \subseteq R$), then the respective sequence is called a sequence of real numbers.

The sequences from the previous examples, 1), 2) and 3), are sequences of real numbers.

From all sequences of real numbers, we are particularly interested in those having a property called *convergence*.

To intuitively understand what a convergent sequence means, without going deeper into the more complex notions of mathematical analysis, we will try to create an image of convergence that can be mentally visualized.

As we saw, a sequence contains an infinite number of terms, arranged in a unique order (corresponding to the enumeration).

Thus, we cannot enumerate or visualize all sequence's terms, but at least a fairly big finite number of terms.

The convergence property of a sequence manifests *at infinity* and it is a matter of the behavior of the terms from a certain rank upward. Therefore, this property cannot be directly observed, but only imagined.

A convergent sequence is a sequence whose terms are accumulating around a finite real number and approaching it as their rank increases. This number is called the limit of the sequence.

So, a sequence is convergent if all its terms from a certain rank upward become closer and closer in value to a constant finite real number.

This number (the limit of the sequence) may be reached (it is one of sequence's terms) or not.

A wording closer to the mathematical definition is the following: a sequence (a_n) is convergent toward limit c ($c \in R$) if, whatever vicinity of c we choose (for example, an interval centered in c), it contains all the sequence's terms starting from a certain rank N ($a_N, a_{N+1}, a_{N+2}, \ldots$). Of course, N depends on the chosen vicinity.

In other words, any vicinity of the limit contains an infinite number of terms of the sequence.

It is written $a_n \to c$ or $\lim_{n \to \infty} a_n = c$.

For the rigorous definitions of notions of vicinity, accumulation and convergence, see the section titled *Sequences of real numbers. Limit.*

Imagining the sequence's terms as points on the real axis, for a convergent sequence these points have a nonexpanded distribution, creating a progressive density around the limit or, in other words, accumulating around that point:

$$\bullet \quad \bullet \quad \bullet \quad \bullet \quad \bullet \quad \bullet \quad \bullet \bullet \quad \bullet\bullet\bullet\bullet ...\circ$$
$$a_0 \ a_1 \ a_2 \ldots\ldots\ldots\ldots a_n \ldots\ldots\ldots\ldots c$$

Here are few examples:

The sequence $\frac{1}{2}, \frac{1}{3}, \frac{1}{4}, \frac{1}{5}, ...$ of real numbers is convergent toward 0 (zero).

Even without a rigorous demonstration, we can observe this sequence's terms decreasing, coming closer and accumulating around zero (without reaching it) as their rank increases. Zero is the limit of this sequence.

Similarly, the sequence $\frac{1}{2}, \frac{1}{4}, \frac{1}{6}, \frac{1}{8}, ..., \frac{1}{2n}, ...$ is convergent toward zero and the sequence $1, \frac{1}{4}, \frac{1}{9}, \frac{1}{16}, ..., \frac{1}{n^2}, ...$ as well.

The sequence $\frac{1}{2}, \frac{2}{3}, \frac{3}{4}, \frac{4}{5}, ..., \frac{n-1}{n}, ...$ is convergent toward 1.

The alternating sequence $1, -1, 1, -1, 1, -1, ...$ is not convergent (we say it is divergent).

To prove this, we can assume the absurd; namely, there exists a finite limit c and show that for any $c \in R$, a vicinity of c exists such that it contains neither 1, nor –1 (all the sequence's terms).

Now, having an intuitive idea of what a convergent sequence means, let us go back to the presentation of probability as a limit.

First we define the basic objects that probability theory operates with.

Experiments, events

The objects upon which probability theory is applied for the purpose of prediction are called events.

In the next section, *Probability as a measure,* we define the events as sets of parts belonging to an ensemble structured as a field with specific properties.

Until then, we use the word *event* having attached its dictionary definition: an event is a possible action, occurrence, process or act with causal determination, which can happen or not, these being the only two options.

An *experiment* is a type of action that generates events.

An example of an experiment is rolling the die. This experiment could be performed by several people, in different places and moments.

The performance of an experiment is called *test*.

Rolling the die, as a procedure, is an experiment, but rolling the die by a certain person, at a certain moment, is a test.

Therefore, a test is the act of putting an experiment into practice in established specific conditions.

The experiments generate *outcomes* (results).

An experiment can have more than one outcome, while a test can have only one outcome.

The experiment of rolling the die can have six outcomes (namely the numbers from 1 to 6 inscribed on die's sides).

These outcomes can occur wherever and whenever this experiment is performed.

But if we perform this experiment at an established moment, it becomes a test and the outcome of respective test can only be one of the six die's numbers.

Probability theory deals with events generated by aleatory experiments; that is, experiments in which the physical conditions of performing do not influence the outcomes from a probability point of view.

An *event* is an arbitrary set of pre-established outcomes of an experiment.

An event can happen (occur) or not, as result of an experiment.

We can enunciate any kind of event related to a given experiment, with the goal of referring to their probability of occurrence. So, the events can be textually predefined.

Corresponding to the experiment of rolling the die, here are some possible or impossible events, denoted by uppercase letters:
A – occurrence of number 2
B – occurrence of number 5
C – occurrence of an even number
D – occurrence of an uneven number
E – occurrence of a less than 3 number
F – occurrence of number 7.

For the experiment of spinning the roulette wheel, here are some events:
A – occurrence of a red number
B – occurrence of number 15
C – occurrence of a less than 12 number
D – occurrence of a black number
E – occurrence of an uneven number
F – occurrence of number 41.

For the experiment of picking a card from a 52-card deck:
A – occurrence of a clubs card
B – occurrence of a diamonds or spades card
C – occurrence of a higher than 10 card
D – occurrence of an ace
E – occurrence of a jack.

If we take the experiment of choosing a group of ten arbitrary people from the crowd, here are some events:
A – 2 persons from the chosen group have the same birthday
B – 3 persons from the chosen group have the same birthday
C – any 2 persons from the chosen group have not the same birthday.

If e is the outcome of a test and A an event related to a respective experiment, we say that A *happens* if $e \in A$ and A *does not happen* if $e \notin A$.

Example: In the test of rolling the die (the first example from the four above), the outcome $e = 3$ occurred (the die showed a 3).

Then, none of events A, B, C and E happen, but event D does happen.

We saw that the events can be represented as sets, so we are allowed to use with them according to the set theory.

In the next section we see that events as sets have additional properties (measurability) when they are represented as elements of a field defined by certain axioms.

Two experiments are said to be *independent* when the conditions of performing the first experiment do not influence the results of the second, and vice versa.

For example, rolling the die by two different people are two independent experiments.

The experiments of consecutive draw of two cards from a deck, without putting back the first card drawn, are not independent.

The relative frequency

In the section titled *Probability – the word*, we defined the probability of event A occurring as being the ratio between the number of tests favorable for A to occur and the number of all equally possible tests.

Let us come back to the classical example experiment of rolling the die.

Let A be the event *occurrence of number 5*.

According to the classical definition, the probability of A is

$P(A) = \dfrac{1}{6}$ (one favorable case; namely, the die shows the side with the number 5 from six equally possible cases).

By establishing the probability $P(A)$, we have attached a unique number to event A (1/6 in this case).

The ratio from the definition of probability is positive and less than 1, so probability is a function P defined on the set of events generated by an experiment, with values in interval $[0,1]$.

This function attaches a positive and less than 1 number to each event and this number is called an event's probability.

The properties of this function, as well as its extension to sets of more complex events, is exposed in the section titled *Probability as a measure*.

The simple assignment of a number to a singular event through function P does not provide additional information about the occurrence of respective event.

The fact that we know the probability of occurrence of number 5 at die rolling as being 1/6 does not confer any relevance to the prediction that this event will happen or not at a given moment as result of a singular test.

But there exists a mathematical result, called the Law of Large Numbers, which gives us additional information about the occurrence of an event within a sequence of experiments.

This information is about the frequency of occurrence of the event within the sequence of tests and is a property of limit making the connection between relative frequency and probability.

Let us see what relative frequency means.

Staying with the example experiment of rolling the die and taking the same event A (occurrence of number 5), let us assume we have a sequence of independent experiments (tests) $E_1, E_2, E_3, ..., E_n, ...$, each generating a certain outcome.

We can choose this sequence as being the chronological sequence of tests of rolling the die, performed by same person over time, or the chronological sequence of such tests performed by whatever number of established persons.

We can also choose the sequence as being the chronological sequence of such tests performed by all the people on Earth who make up this experiment.

No matter the set of chosen tests, as long as they are well defined and form a sequence (an infinite enumeration can be attached to them).

Obviously, any of these choices is hypothetical.

Within this sequence of independent tests $(E_n)_{n \geq 1}$, we define the relative frequency of occurrence (producing) of event A.

Let us assume the die rolls a 2 on the first test (E_1). At this moment (after one test), the number of occurrences of 5 is 0.

Denote: $E_1 - 1 - 0$

On the second test (E_2), the die rolls a 1. At this moment (after two tests), the number of occurrences of 5 is still 0.

Denote: $E_2 - 2 - 0$

On the third test (E_3), the die rolls a 4. At this moment (after three tests), the number of occurrences of 5 is still 0.

Denote: $E_3 - 3 - 0$

On the fourth test (E_4), the die rolls a 5. At this moment (after four tests), the number of occurrences of 5 is 1.

Denote: $E_4 - 4 - 1$

On the fifth test (E_5), the die rolls a 2. At this moment (after five tests), the number of occurrences of 5 is still 1.

Denote: $E_5 - 5 - 1$

And so on, assume we obtain the following results:

$$E_1 - 1 - 0$$
$$E_2 - 2 - 0$$
$$E_3 - 3 - 0$$
$$E_4 - 4 - 1$$
$$E_5 - 5 - 1$$
$$E_6 - 6 - 1$$
$$E_7 - 7 - 1$$
$$E_8 - 8 - 1$$
$$E_9 - 9 - 2$$
$$E_{10} - 10 - 2$$
$$............$$
$$E_n - n - a_n$$
$$............$$

In the diagram above, the first column contains the successive tests, the second column contains the order numbers within the sequence and the third column contains the cumulative numbers of occurrences of 5.

The successive results from the third column (the total number of occurrences of 5 after each test) form a sequence $(a_n)_{n\geq 1}$ (observe in the diagram that they are values of a function defined on N; namely, an infinite enumeration).

Thus, we have obtained the sequence 0, 0, 0, 1, 1, 1, 1, 1, 2, 2,..., a_n, ... of cumulative numbers of occurrences of event A as the result of the performed tests.

Probability theory does not provide us with any property of this sequence $(a_n)_{n\geq 1}$, so we have no information about certain of its terms, nor about its behavior to infinity.

What we can only observe is the sequence being monotonic increasing (see the section titled *Sequences of real numbers. Limit* in the mathematical chapter).

In exchange, the theory provides information about another sequence, namely the sequence $\left(\dfrac{a_n}{n}\right)_{n\geq 1}$.

By making the ratio a_n/n for each test, we can complete the previous diagram with a new column containing the numerical values of these ratios:

$$E_1 - 1 - 0 - 0$$
$$E_2 - 2 - 0 - 0$$
$$E_3 - 3 - 0 - 0$$
$$E_4 - 4 - 1 - 1/4$$
$$E_5 - 5 - 1 - 1/5$$
$$E_6 - 6 - 1 - 1/6$$
$$E_7 - 7 - 1 - 1/7$$
$$E_8 - 8 - 1 - 1/8$$
$$E_9 - 9 - 2 - 2/9$$
$$E_{10} - 10 - 2 - 2/10 \,(= 1/5)$$

....

$$E_n - n - a_n - a_n/n$$

....

48

Obviously, the results from the last column form a sequence, as being values of a function defined on N (0, 0, 0, 1/4, 1/5, 1/6, 1/7, 1/8, 2/9, 2/10,..., a_n/n, ...).

The general term of this sequence is $\dfrac{a_n}{n}$ and is called the *relative frequency* of occurrence of event A.

Therefore, the relative frequency is the ratio between the number of tests having as a result the occurrence of event A and the total number of tests performed.

The result offered by probability theory with respect to the sequence of relative frequencies is a theorem known as the Law of Large Numbers.

This theorem asserts that *for any sequence of independent experiments and any event* A, *the relative frequency of occurrence of event* A *converges toward the probability of event* A.

We may write this as $\lim\limits_{n \to \infty}\left(\dfrac{a_n}{n}\right) = P(A).$

For our example, the theorem states that the sequence of real positive numbers 0, 0, 0, 1/4, 1/5, 1/6, 1/7, 1/8, 2/9, 2/10, ..., a_n/n, ... is convergent and its limit is $P(A)$, namely 1/6.

That means, as n is increasing, the terms a_n/n are getting closer and closer to 1/6 and are accumulating around this number, approaching it by any decimal places from a certain rank n upward.

Obviously, the conclusion of the theorem is valid for any sequence of independent tests and any chosen event.

The information provided by the Law of Large Numbers with regard to the sequence $\left(\dfrac{a_n}{n}\right)$ is a limit one.

We have no information about certain terms $\dfrac{a_n}{n}$, nor about the occurrence of event A after a certain finite number of tests, but we know only how these terms behave to infinity (namely, they are approaching the probability of event A).

In any sequence of tests $(E_n)_{n \geq 1}$ we choose, the sequence of relative frequencies $\left(\dfrac{a_n}{n}\right)_{n \geq 1}$ will be convergent toward limit $P(A)$.

According to a property of convergent sequences, any subsequence of the sequence of tests $(E_n)_{n \geq 1}$ will be convergent toward P(A), too (a subsequence is obtained from a sequence by successively choosing terms whose indexes form an increasing sequence of natural numbers; for example, $E_2, E_4, E_6, E_8, ..., E_{2n}, ...$ is a subsequence of the sequence $(E_n)_{n \geq 1}$).

The Law of Large Numbers grants the numerical probability with a limit property.

The probability of an event (calculated after the classical definition) is the limit of a sequence of relative frequencies generated by a sequence of independent experiments.

Moreover, this is a special limit because it is the limit of every sequence of relative frequencies: no matter what sequence of independent experiments we may choose, the associated relative frequencies (of the occurrence of event A) are convergent toward the same limit, which is the probability of A.

Probability as a measure

In the previous section we presented probability as a number associated with an event generated by an experiment with a finite number of outcomes.

The probability was defined particularly, on a finite set of events, in which the elementary events are equally possible.

The notions of elementary event, field of events and equally possible events were not defined rigorously, but only through examples.

In this context, the probability of an event is defined as the ratio between the number of cases that are favorable for the event to occur and the total number of equally possible cases.

By following this definition, we can easily calculate the probabilities of events generated by experiments like rolling the die, spinning the roulette wheel, picking a card from a given deck, drawing an object from an urn whose content is known, etc.

We can apply the classical definition in these cases because each of the experiments enumerated generates a finite number of events and the elementary events are equally possible.

The classical definition of probability does not refer to the set of events associated with an experiment, which may be more or less complex, but only to its number of elements.

Once this set is organized (structured) as a field with certain axioms (properties), the probability can be defined as a function whose properties are generated by the structure of the respective field and can be studied in greater depth.

In mathematics, a structure is assumed to organize a set of objects (which could be numbers, points, sets, functions, and the like) by introducing a group of axioms enunciating properties of the elements with respect to certain relations of operations (composition laws) that are well defined for that set.

By granting a set a well-defined structure, the set acquires the denomination of field, space, corpus, algebra, and so on.

In theory, organizing sets as structures is generally done to study the properties of certain functions that are defined on those sets.

Here are few examples of structures:

1) The group structure (in algebra)

A set G along with a law of internal composition (that is a function " $*$ " defined on the Cartesian product $G \times G$ with values in G, like, for example, the addition or the multiplication in the real system R) is called *group* if the following axioms are met:

a) $\forall a, b, c \in G$, then $a * (b * c) = (a * b) * c$ (associativity)

b) $\forall a \in G, \exists e \in G$ such that $a * e = e * a = a$ (existence of a zero element)

c) $\forall a \in G, \exists a' \in G$ such that $a * a' = a' * a = e$ (every element has a reciprocal element).

The set G granted with the composition law " $*$ " is denoted by $(G, *)$.

Some examples of groups are the group of real numbers (with respect to addition operation) – $(R, +)$; the group of positive rational numbers (with respect to the multiplication operation) – $(Q_{+}, \cdot)$; and the group of $n \times n$ – size matrixes (with respect to the operation of addition of matrixes) – $(\mathcal{M}_n, +)$.

Granting the set R with the addition operation "+" confers on it a group structure.

The *set R* and the *algebraic group R* are two different objects.

2) The topological space structure (in mathematical analysis, topology)

If X is a non-empty set, call *topology* on X a family of subsets of X, denoted by τ, which meets the following three axioms:

a) $\phi, X \in \tau$;

b) $D_1, D_2 \in \tau$ implies $D_1 \cap D_2 \in \tau$ (it is closed to finite intersection);

c) If $D_i \in \tau$ for any $i \in I$, then $\bigcup_{i \in I} D_i \in \tau$ (it is closed to arbitrary union).

The couple (X, τ) is called *topological space*.

By defining a topology on an arbitrary set, that set acquires a topological space structure.

An example of topological space is the set R along with the family of open intervals of real numbers.

3) The total order structure (in algebra, mathematical analysis)

A relation $\mathcal{R}$ between the elements of set A (a relation is a correspondence established between the elements of a set) is said to be a *total order relation* if the following axioms are met:

a) $\forall a, b \in A$, then $a \mathcal{R} b$ or $b \mathcal{R} a$ (any two elements are comparable).

b) $\forall a, b, c \in A$ such that $a \mathcal{R} b$ and $b \mathcal{R} c$, then $a \mathcal{R} c$ (transitivity).

A set on which we have defined a total order relation is called fully ordered (it acquires a total order structure).

For example, set R along with the well-known relation $\leq$ is a fully ordered set.

The set of parts of a set A, namely $\mathcal{P}(A)$, along with relation of inclusion between sets $\subset$, is not fully ordered (relation $\subset$ meets the transitivity, but does not meet the first axiom).

These are examples of mathematical structures defined through groups of axioms.

Structures are defined and attached to sets to generate new concepts. A set granted a certain structure is a mathematical object that is different from the initial set, even if the language of communication uses the same words many times to express them.

For example, set R^2 (the coordinate plane) may be structured as an algebraic group (by attaching a composition law meeting the group's axioms), as a topological space (by attaching a topology), as a metric space (by attaching the classical orthogonal distance), as a vector space, as an algebraic ring, etc.

Coming back to the set of events generated by an experiment, we will give it specific structure in order to define a function on this set that extends the classical definition of probability.

In the previous section we saw that events can be represented as sets (of outcomes) and the classical operations of union and intersection can work between them.

Staying with the finite case, namely at the finite sets of events generated by an experiment:

If Ω stands for the set of all possible outcomes of an experiment, and $\mathcal{P}(\Omega)$ is the set of all parts of Ω, then the events are elements of $\mathcal{P}(\Omega)$.

For example, the possible outcomes of the experiment of rolling the die form the set $\Omega = \{1, 2, 3, 4, 5, 6\}$. Events like $\{1\}$, $\{2, 3\}$, $\{3, 4, 5\}$ are elements of $\mathcal{P}(\Omega)$.

Denoting by Σ the set of the events associated with an experiment having Ω as set of all possible outcomes, generally we have $\Sigma \subset \mathcal{P}(\Omega)$.

On the set Σ of events associated with an experiment we can introduce (define) three operations corresponding to the logical operators *or, and* and *non*.

Thus, given two arbitrary events $A, B \in \Sigma$, we can define:

– The event *A or B* as being the event that occurs if and only if at least one of events *A, B* occurs; it is denoted by $A \cup B$;

– The event *A and B* as being the event that occurs if and only if both of events *A, B* occur; it is denoted by $A \cap B$;

– The event *non A*, as being the event that occurs if and only if event *A* does not occur; it is called the opposite of *A* and it is denoted by A^C.

The logical operations between events defined above correspond to the operations between sets, respectively union, intersection and complement. The events can be represented as sets, so the denotations are justified.

In the set of events Σ exist two events with special significance; namely, the event $\Omega = A \cup A^C$ and the event $\phi = A \cap A^C$.

The first consists of the occurrence of event A or the occurrence of its opposite, that always happens. It is then natural to call it the *sure event*.

The second consists of the occurrence of event A and the occurrence of its opposite, that is impossible. It is then natural to call it the *impossible event*.

Observe that events Ω and ϕ do not depend on A.

We have used thus far the words *elementary event* without attaching a rigorous definition.

We first define the compound event:

An event A is said to be *compound* if it can be written as the union of two events, each being different from A ($\exists B, C \in \Sigma$, $B \neq A$, $C \neq A$, such that $A = B \cup C$).

If event A is not compound, it is said to be *elementary*.

As examples for the experiment of rolling the die:

– Event $\{3, 5\}$ is compound, because $\{3, 5\} = \{3\} \cup \{5\}$;

– Event $\{1, 2, 4\}$ is compound, because $\{1, 2, 4\} = \{1, 2\} \cup \{4\}$;

– Events $\{1\}, \{2\}, \{3\}, \{4\}, \{5\}, \{6\}$ are elementary.

Two events A and B are said to be *incompatible* or *exclusive* if $A \cap B = \phi$, that is, A and B cannot occur simultaneously.

In the experiment from our example, events $\{3, 5\}$ and $\{1, 2, 4\}$ are incompatible.

As we said, we will attach a structure to the set of events Σ that allows us to define a certain function and to study its properties.

It is about Boole algebra, which assumes granting a set with some operations between its elements, operations that meet a certain group of axioms.

Boole algebras have immediate application in logic and informatics, but also in mathematical analysis and measure theory.

A Boole algebra is a non-empty set $\mathcal{A}$, in which three logical operations are defined *(or, and, non)* and these operations meet a group of five axioms: commutativity and associativity (for each of operations *or* and *and*), absorption (for each of operations *or* and *and*), distributivity (of operation *and* given *or* and reverse), complementarity (absorption of an element operated with its opposite).

See the explicit forms of these axioms in the section titled *Boole Algebras* in the mathematical chapter.

There are Boole algebras of sets, sentences, equivalence classes, and the like.

The simplest example of Boole algebra is the set of parts $\mathcal{P}(\Omega)$ of a non-empty set Ω, along with the operations of union, intersection and complement (with respect to Ω) between sets.

You can see these operations of the given set meeting all the five axioms of the definition.

This is also the reason for denoting the three Boolean operations by $\cup$ (or), $\cap$ (and) and ^{C}, even if algebra's elements are not always sets (also the symbols $\vee$, $\wedge$ and $^{-}$ of the logical operators are used).

From the definition of a Boole algebra come several consequences that enunciate properties of this structure. You can also run through these results in the same *Boole Algebras* section in the mathematical chapter.

We underline here two of these consequences:

Consequence 4: For any sets $(A_i)_{1 \le i \le n} \subset \mathcal{A}$, the elements $A_1 \cup \ldots \cup A_n$ and $A_1 \cap \ldots \cap A_n$ are uniquely determined and do not depend on the order of operated elements.

Because the set of parts $\mathcal{P}(\Omega)$ is a Boole algebra, the union and intersection of sets have the properties stated in this consequence; therefore, the denotations $A_1 \cup \ldots \cup A_n$ and $A_1 \cap \ldots \cap A_n$ (having no brackets) are justified, as well as notions of finite union and finite intersection of sets.

Consequence 5: In a Boole algebra two elements exist, called the null element, denoted by Λ, and the total element, denoted by V, so that for any $A \in \mathcal{A}$, the following equalities are true:
$$A \cap A^{C} = \Lambda \text{ and } A \cup A^{C} = V.$$

In the algebra $\mathcal{P}(\Omega)$, we have $V = \Omega$ and $\Lambda = \phi$.

We also have a definition of minimal elements with respect to the relation order $\subset$ (the equivalent of the implication relation $\Rightarrow$ from logics), called *atoms:*

An element A of a Boole algebra $\mathcal{A}$, $A \neq V$ is called atom of that algebra, if the inclusion $B \subset A$ implies $B = \Lambda$ or $B = A$, for any $B \in \mathcal{A}$.

In the algebra $\mathcal{P}(\Omega)$, each part of Ω having one single element is an atom of this algebra.

The set Σ of the events associated with an experiment, along with previously defined operations between events, form a Boole algebra.

This result can be immediately deduced if it is taken into account that the events can be represented as sets and Σ is included in $\mathcal{P}(\Omega)$, but it also can be stated as an axiom if, from a rigorousness excess, we do not identify the event with the set of tests that generate it.

Between the specific notions of a Boole algebra and those of the set of events associated with an experiment, we can observe the following correspondence:

Boole algebra	The set of events (Σ)
operation $\cup$	operation *or*
operation $\cap$	operation *and*
operation ^{c}	operation *non*
null element Λ	impossible event (ϕ)
total element V	sure event (Ω)
atom	elementary event

The Boole algebra of the events associated with an experiment is called the *field of events* of the respective experiment.

A field of events is then a set of results Ω, structured with an algebra of events Σ and is denoted by $\{\Omega, \Sigma\}$.

Thus, we have attached an algebraic structure to each experiment; namely, the Boole algebra of the set of events Σ.

This action creates the basis of the mathematical model on which the real phenomenon can be studied, making possible the step from the practical experiment to probability theory.

Granting the set Σ with an algebraic structure has the goal of conferring consistency to the ulterior definition of probability as a function on Σ and as well of providing us the tools needed to deduce the properties of this function.

These properties stand for the basic formulas of applied probability calculus.

The first step in extending the classical definition of probability is defining it as a function of a finite field of events.

A field of events $\{\Omega, \Sigma\}$ is finite if the total set Ω is finite.

The next definition calls probability a function on a finite field of events, which has three certain properties.

Let $\{\Omega, \Sigma\}$ be a finite field of events.

Definition: Call probability on Σ, a function P: $\Sigma \rightarrow R$ meeting the following conditions:
   (1) $P(A) \geq 0$, for any $A \in \Sigma$;
   (2) $P(\Omega) = 1$;
   (3) $P(A_1 \cup A_2) = P(A_1) + P(A_2)$, for any $A_1, A_2 \in \Sigma$ with $A_1 \cap A_2 = \phi$.

From this definition, it follows that:
   1) A probability takes only positive values;
   2) The probability of the sure event is 1;
   3) The probability of a compound event consisting of two incompatible events is the sum of the probabilities of those two events.

Probability is defined then as a function P on the field of events associated with an experiment, which meets the three conditions (axioms) described above.

The fact that $\{\Omega, \Sigma\}$ is a field of events (that reverts to the fact that Σ is structured as a Boole algebra) ensures:

– The membership $\Omega \in \Sigma$ (therefore the expression $P(\Omega)$ from condition (2) does make sense);

– The commutativity of the operation of union (*or*) between events and the membership $A_1 \cap A_2 \in \Sigma$ (therefore the expression $P(A_1 \cup A_2)$ from condition (3) does make sense).

This information ensures the total consistency of the definition.

Property (3) can be generalized by recurrence for any finite number of mutually exclusive events. Therefore, if $A_i \cap A_j = \phi$, $i \neq j$, $i, j = 1, \ldots, n$, then:

$P(A_1 \cup A_2 \cup \ldots \cup A_n) = P(A_1) + P(A_2) + \ldots + P(A_n)$ or, else

written, $P\left(\bigcup_{i=1}^{n} A_i\right) = \sum_{i=1}^{n} P(A_i)$.

The axioms of a Boole algebra (commutativity and associativity) are also involved here, allowing the above denotations by defining in a solid way to operate a *finite union* within an algebra of events.

This is the definition of probability in a finite field of events. Let us name it *Definition 1*.

It does not identify a unique function, called probability, but it attaches this name to a function defined in the field of events, which has the three specific properties.

It is then possible for several probabilities (as functions) to exist in the same field of events, all having the same properties presented in the mathematical chapter.

Definition 1 does make sense as long as the set of events Σ has a Boolean structure (it is a field of events).

Therefore, we cannot talk about the probability of an isolated, singular, predefined event. We can do that only in the context of its membership in a field of events.

Thus, probability makes sense only as a function of a well-defined set that is structured as a field of events.

A finite field of events $\{\Omega, \Sigma\}$, along with a probability P, is called a *finite probability field* and is denoted by $\{\Omega, \Sigma, P\}$.

Two properties of the elementary events from a finite algebra of events say that every event can be written in a unique form as a union of elementary events, and the sure event is the union of all elementary events (properties (E6) and (E7) from the section titled *Field of events* in the mathematical chapter).

By using these properties and axiom (3) from Definition 1, we deduce that, to know the probabilities of all events from Σ, is enough to know the probabilities of the elementary events that form the total set $\Omega = \{\omega_1, \omega_2, \ldots, \omega_r\}$ (let these be $P(\{\omega_i\}) = p_i$, $1 \leq i \leq r$).

The demonstration is very simple and can be followed in the section titled *Probability on a finite field of events* in the mathematical chapter.

This result tells us that a finite probability field is completely characterized by the non-negative numbers $p_1, p_2, ..., p_r$, with

$$\sum_{i=1}^{r} p_i = 1.$$

If the probabilities of the elementary events are equal, we find the classical definition of probability, which automatically becomes a particular case of Definition 1.

From Definition 1 and the properties of a Boole algebra come the most important properties of probability, namely the properties (P1) – (P10) from the section titled *Probability on a finite field of events* in the mathematical chapter.

Thus far, we have defined probability as a function with certain properties on a finite field of events.

Even from Definition 1 we can observe a measurability feature of probability resulting from axiom 3, which is a partial additivity property: in certain conditions (incompatible events), the value of function P in a compound event is the sum of the values of P in the two constituent events.

This property is expressed in common language through the fact that probability measures the events, the mathematical concepts of *measure* and *measurability* being rigorously defined in the mathematical chapter (in the section titled *Measure Theory Basics).*

This is in fact one of this chapter's goals, to present probability as a measure.

To do that, we will rebuild the mathematical definition of measure by its constituent notions and identify the probability with this new definition.

Let us go back to Definition 1 of probability. To generalize this definition, we must weaken the conditions from the hypothesis; namely, we must define the function P on a set having a more complex structure than a finite field of events.

The immediate generalization is obtained by considering the total set Ω as an arbitrary set and Σ an algebra of events that is closed not only to finite union, but also to *countable union.*

A set A is called countable if a bijective function exists from A to N (the set of natural numbers). Thus, a unique natural number is associated with each element of A, and the reverse, thus justifying the term *countable,* with total coverage in the common language (countable = that which can be counted).

Obviously, a countable set is infinite (because N is infinite and that function is bijective).

The simplest example of a countable set is even $N = \{0, 1, 2, 3, ...\}$.

Other countable sets are: $\{0, 2, 4, 6, 8, ...\}$ – the set of even natural numbers; $\left\{ \dfrac{1}{n}, n \in N^* \right\}$ – the set of fractions of $\dfrac{1}{n}$ form; the set Q of rational numbers; $\{ [n, n+1], n \in N \}$ – the set of closed intervals of $[n, n+1]$ form, etc.

Returning to the countable union, this term extends the finite union ($A_1 \cup A_2 \cup ... \cup A_n$), consistently defined by Boole algebra's axioms, to the union of a countable family of sets ($\bigcup_{i \in I} A_i$, where I is a countable set).

The complete mathematical definition of a union of arbitrary (infinite) family of elements from a Boole algebra appears in the section titled *Tribes. Borel sets. Measurable space* in the mathematical chapter.

This definition also stands in particular for the countable union (corresponding to a countable family). For this definition of countable union to be consistent, it is necessary to modify the axioms of a Boole algebra by adding the property of being closed to this operation: for every countable family of elements of the algebra, their union is still an element of the algebra.

Thus, we have built the definition of a new concept, called σ-algebra or tribe: a σ-algebra is a Boole algebra that is closed to the countable union or, in other words, has the countable additivity property (a Boole algebra $\mathcal{T}$ is a tribe if $A_n \in \mathcal{T}, n = 1, 2, 3, ...$

$$\Rightarrow \bigcup_{n=1}^{\infty} A_n \in \mathcal{T}) .$$

From the properties of a Boole algebra immediately results that a tribe is also closed to the countable intersection

$$(A_n \in \mathcal{T}, n = 1, 2, 3, \ldots \Rightarrow \bigcap_{n=1}^{\infty} A_n \in \mathcal{T}).$$

A field of events having the countable additivity property (that is, a tribe) is called σ-field of events.

The new definition of probability uses a set of events structured as a σ-algebra for the domain of function P and replaces the finite additivity axiom with one of countable additivity:

Definition: Let $\{\Omega, \Sigma\}$ be a σ-field of events. Call probability on $\{\Omega, \Sigma\}$ field a function P: $\Sigma \to R$ obeying the following conditions:

(1) $P(A) \geq 0$, for any $A \in \Sigma$ (non-negativity);

(2) $P(\Omega) = 1$ (normalization);

(3') $P\left(\bigcup_{i=1}^{\infty} A_i\right) = \sum_{i=1}^{\infty} P(A_i)$, for any countable family of mutually

exclusive events $(A_i)_{i=1}^{\infty} \subset \Sigma$ (countable additivity).

Let us call it *Definition 2*.

A σ-field of events $\{\Omega, \Sigma\}$ along with a probability P is called probability σ-field and is written as $\{\Omega, \Sigma, P\}$.

The left-hand member of axiom (3') contains the symbol for a countable union, defined previously, and the right-hand member contains the symbol $\sum_{1}^{\infty}$ for an infinite sum.

We will make a parenthesis to define this symbol.

Let us consider an arbitrary sequence of real numbers $(a_n)_{n \geq 1}$ and the partial sums of its terms:

$s_1 = a_1$

$s_2 = a_1 + a_2$

$s_3 = a_1 + a_2 + a_3$

$\ldots\ldots\ldots\ldots\ldots$

$s_n = a_1 + a_2 + a_3 + \ldots + a_n .$

The numbers s_n are well defined (they are finite sums) and form in turn a sequence $(s_n)_{n \geq 1}$, called the sequence of partial sums.

The couple of sequences $((a_n)_{n \geq 1}, (s_n)_{n \geq 1})$ is called a *series* of numbers.

If the sequence of partial sums $(s_n)_{n \geq 1}$ is divergent, then the respective series is said to be divergent.

If the sequence of partial sums has a limit, then this limit is denoted by $\sum_{n=1}^{\infty} a_n$ and is called the *sum of the series*.

If this limit is a finite number, the respective series is said to be convergent.

So, the right-hand member from axiom (3') of Definition 2 (the infinite sum) is the sum of a series, respectively, the limit of the sequence of partial sums $(s_n)_{n \geq 1}$, where

$s_n = P(A_1) + P(A_2) + ... + P(A_n)$.

Axiom (3') assumes not only the numerical equality of the two members, but also the fact that the respective series is convergent (because the left-hand member is a finite number) for that summation symbol to make sense.

Let us go back to Definition 2 as a whole.

This axiomatization of probability belongs to Kolmogorov, who states that infinite probability fields are idealized models of real random processes, and that he limits himself arbitrarily to only those models that satisfy countable additivity.

This axiom (3') is the cornerstone of the assimilation of probability theory into measure theory.

There are other axiomatizations that give up normalization, and that give up countable additivity and even additivity.

In this work, when we speak of probability and probability calculus, we mean Kolmogorov's approach—recognized as the standard—and we use Definition 2 as the complete definition of probability.

Further, we see what a measure means in mathematics and how the properties of probability make it a particular measure.

The word measure is used in common language when referring to a numerical value that is attached to an object as result of a measurement upon it.

We measure the length of a distance, the mass of a corpus or the volume of a liquid, in order to attach to these objects numerical values as the result of the measurements.

Even if for the act of measuring we use specific tools or devices, this action is in fact a multiplication of a standard measure (the fixed measure of a pre-established object from the same category).

For example, if the result of measuring the length of a segment is seven meters, this means that the 1-meter measure (the fixed length of a pre-established standard segment) multiplied seven times fits exactly the length of the measured segment, or, in other words, the segment consists of seven equal segments, each equal to one standard segment.

Thus, the result of a measurement is not an absolute numerical value, but a multiple (real number) of a chosen standard measure.

This example shows that the act of measurement has a relative quantitative characteristic that preserves the order (the bigger the objects, the higher their measures).

To create a mathematical model of measurement that captures these properties, it is first necessary to take the measure as a function with non-negative numerical values, and the measurable objects as sets organized through a certain structure.

This function has to be monotonic increasing (to preserve the order), with respect to the order relation established between the objects to be measured (respectively, the inclusion relationship in case of sets).

Therefore, the act of measurement becomes a function on the set of objects to be measured, and the measure of an object is the value of that function for the respective object.

The set of measurable objects (the domain of the measurement function) must have a mathematical structure that allows the operation of two or more objects and must define an order relation.

In case we take the objects to measure as sets, the domain's structure can be a Boolean one (algebra of sets).

Coming back to the example of measuring length, let us imagine the domain as the set of all simple or compound segments (we named a compound segment any segment that is a union of line segments, respectively any polygonal path).

The length is then a function L defined on this domain, taking values in interval $[0, \infty]$, which associates with each segment the real number resulting from multiplication of the fixed standard length in that segment.

For two segments a and b having no common points, we have $L(a \cup b) = L(a) + L(b)$.

This is a finite additivity property of function L.

To be valid, the mathematical model of measure must have this property or a more general one.

On the real axis R, defining the length of an interval as the difference between its abscissas is a small example of building a measure on the set of intervals of real numbers.

The mathematical effort of consistently defining a measure that extends the notion of length of an interval for more complex sets was done by G. Cantor, who continued the development of measure theory started by E. Borel.

We do not present in this chapter all notions that are the ground for measure theory—these can be viewed in the section titled *Measure Theory Basics* in the mathematical chapter.

We limit the discussion instead to the general definition of measure and few of its properties in order to make the connection with probability theory.

As we stated earlier, building an ideal model for the measure assumes a function (the measure) having certain properties to be defined on a set of sets with a specific structure.

This definition space of the measurable objects (the function's domain along with its mathematical structure) is essential to the consistency of the mathematical model.

The structure chosen by Cantor to stand for the domain of the measure function is the σ-algebra of sets (tribe), which we have already presented in this chapter.

It is a Boole algebra having sets as elements and, in addition, the countable additivity property (the union of any countable family of sets is still an element of that algebra).

If X is a set and $\mathcal{T}$ is a tribe on X, then the couple $(X, \mathcal{T})$ is called *measurable space*, and the elements of $\mathcal{T}$ are called *measurable sets*.

A measure function can only measure measurable sets; that is, sets that belong to a space that is structured as a tribe of parts.

The general definition of measure follows:

Definition: Let $(X, \mathcal{T})$ be a measurable space. Call *measure* on X, a function $\mu : \mathcal{T} \to [0, \infty]$ that has the following properties:

1) μ is a countable additive (or σ-additive) that means: for any sequence $(E_i)_{i \in \mathbb{N}}$ of elements from $\mathcal{T}$, mutually exclusive, we have:

$$\mu\left(\bigcup_{i=1}^{\infty} E_i\right) = \sum_{i=1}^{\infty} \mu(E_i);$$

2) There exists at least one set $E_0 \in \mathcal{T}$ such that $\mu(E_0) < \infty$.

The triplet $(X, \mathcal{T}, \mu)$ is called a *measure space*.

The fact that μ is defined within a tribe structure ensures the consistency of the denotations from axiom 1).

This definition stands for the mathematical model of well-known measures from daily life like length, mass, volume and area.

You can read about other examples of measures in the corresponding section in the mathematical chapter.

From the definition of measure and the axioms of a σ-algebra result all properties of this special function (finite additivity, monotony, countable subadditivity, convergence properties), which you can follow in the mathematical chapter.

Now, if we pay attention in parallel to the definition of measure and Definition 2 of probability, we observe that they are almost identical:

– The definition domains of functions P and μ are structured as tribes;

– Functions P and μ only take non-negative values;

– Each of the functions P and μ has the countable additivity property.

The only conditions that are different are $P(\Omega) = 1$ (for function P) and *There exists at least one set* $E_0 \in \mathcal{T}$ *such that* $\mu(E_0) < \infty$ *(for function μ).*

But if $P(\Omega) = 1$, by using probability's properties, results
$P(\phi) = P(\Omega^C) = 1 - P(\Omega) = 1 - 1 = 0 < \infty$.

Hence, we found a set $E_0 \in \Sigma$ (namely the empty set), with $P(E_0) < \infty$.

This means that function P also obeys axiom 2) of the definition of measure, and the immediate conclusion is that P is a measure.

Therefore, *probability is a measure* on a σ-field of events.

On the other hand, not every measure is a probability.

Now, we can rephrase the definition of probability as follows:

Definition 3: Call probability a measure P defined on a σ-field of events $\{\Omega, \Sigma\}$, with $P(\Omega) = 1$.

Being a measure, the probability will acquire all its specific properties, as they are presented in the mathematical chapter.

Among them, monotony: if $A \subseteq B$, then $P(A) \le P(B)$.

Because for any $A \in \Sigma$ we have $A \subseteq \Omega$, according to monotony will result in $P(A) \le P(\Omega) = 1$, for any event A, and this means that the values of P are less than or equal to 1.

Probability is then a function that measures events, by assigning them values from the interval $[0,1]$.

We tried in this chapter to explain in detail what probability means, especially when referring to the qualitative aspect of this notion.

As we said before, any general explanation of the term *probability* cannot exclude its strictly mathematical definition.

This is why the exposition frequently points to definitions and results from the mathematical chapter.

Obviously, readers having a mathematical background can follow the text much easier than others, the comprehension of mathematical notions and connections between them being more accessible for them.

The goal of this chapter is to create a clear, complete, overall image of the probability concept and to provide the readers with a mathematical analysis, for them to catch the basic qualitative aspect of the notions presented.

Even though the declared goal of this guide is to help readers calculate probabilities, we considered an axiomatic and even philosophical initial approach absolutely necessary because erroneous qualitative interpretations can easily lead to errors in application and calculation.

Also, choosing probability as a decision-making criterion in daily life must be based on a more profound theoretical knowledge beyond an unconditioned reference to statistics.

Readers with a minimal or precarious mathematical background can build themselves a general image of the concept, even without understanding all the mathematical notions and deductions, by simply being conscious that this concept cannot be isolated from a certain axiomatic context, which inevitably is relative and ideal.

This is also the reason we recommend that these people read or re-read this chapter before following the probability calculus guide, even though this last one can be used algorithmically, without going deeper into the notions involved in calculus formulas.

As a whole, the following ideas must be kept in mind:

- Probability is nothing more than a measure; as length measures distance and area measures surface, probability measures aleatory events. As a measure, probability is in fact a function with certain properties, defined on the field of events generated by an experiment.
- A probability is characterized not only by the specific function P, but by the entire aggregate *the set of possible outcomes of the experiment – the field of generated events – function P,* called probability field; probability makes no sense and cannot be calculated unless we initially rigorously define the probability field in which we operate.
- Probability is not a punctual numerical value; textually given an event, we cannot calculate its probability without including it in a more complex field of events. Probability as a number is in fact a limit, respectively the limit of the sequence of relative frequencies of occurrences of the event to measure, within a sequence of independent experiments.

The understanding of probability is complete when it hints not only at the concept's definition, but also at the relationship between the mathematical model and the real world of random processes.

In this direction, the next sections of this chapter are useful for completing the general image created thus far by defining probability.

Relativity of probability

When we speak about the relativity of probability, we refer to the real objective way in which probability theory models the hazard and in which the human degree of belief in the occurrence of various events is sufficiently theoretically justified to make decisions.

Thus, any criticism of the application of probability results in daily life will not hint at the mathematical theory itself, but at the transfer of theoretical information from the model to the surrounding reality.

Probability theory was born from humans' natural tendency to predict happenings and the unknown.

The starting point was the basic experimental observation of the behavior of relative frequencies of occurrence of events within the same type of experiment: according to long-time experience, the relative frequency of occurrence of a certain event oscillates around a certain value, approximating it with high enough accuracy after a very large number of tests.

Aiming to demonstrate this result, probability theory was created step by step and integrated with measure theory, and the experimentally observed property of relative frequency was theoretically proved and called the Law of Large Numbers.

Randomness

Apparently paradoxically, as probability theory developed, the terms *hazard* and *randomness* disappeared from its language, even though they represented the reality object that initially stimulated its creation.

The mathematical model created for probability, which started from hazard and randomness, is exclusively dedicated to defining the measure–function that reproduces the probability behavior of events within a structure, as well as to the deduction of the properties of this function and its statistical applications.

Randomness, as the object of objective reality, is not defined or introduced in the mathematical context of theory. At most, the introductions to the various papers in this domain only make reference to it.

Although the philosophy of hazard has always stood for an attractive field, we note that no one from among the great philosophers has studied the hazard as a philosophical object.

In Kantian language, the hazard does not stand for the thinker as a logical category, nor as an *a priori* form, nor even as a precise experimental category, remaining at its simple status of a word that covers unclassified circumstantial situations of the various theories.

Besides philosophy, mathematics did not succeed in providing a rigorous definition; moreover, it did not create a solid model for randomness and hazard.

Emile Borel stated that, unlike other objects from the surrounding reality for which the creation of models assumes an idealization that preserves their properties, this idealization is not possible in the case of hazard.

In particular, whatever the definition of a sequence formed by the symbols 0 and 1 is, this sequence will never have all the properties of a sequence created *at random*, except if it is experimentally obtained (for example, by tossing a coin in succession and putting down 0 if the coin shows heads and 1 if the coin shows tails).

Borel also proposes an inductive demonstration scheme for this affirmation regarding the *random sequence*.

Assume someone is building such an indefinite sequence 0010111001..., which has all properties of the sequences generated by experimental randomness.

Assume the first n terms of the sequence were built and follows to write the $n + 1$ term.

There are two options: the first n results are somehow taken into account or they are not.

The first option annuls the random character of the construction, because a precise rule for choosing that term exists.

The second option brings us to the same situation we would stand in at the beginning of the sequence's construction: How do we choose one of the symbols 0 or 1 without taking into account any difference between these?

This choice would be equivalent to a draw, which assumes an experimental intervention.

Using the reduction to the absurd method, we come to the conclusion that such a random sequence cannot be built.

Far from entirely convincing us, this proof underlines the theoretical difficulties of conception with regard to this subject.

Borel asserts that we cannot state a constructive definition of a random sequence, but we can attach to these words an axiomatic–descriptive definition.

However, he also admits that such a definition would have no mathematical effectiveness with respect to its integration into probability theory.

In the attempt to mathematize randomness, Richard von Mises has tried to enunciate an axiomatic definition by introducing the notion of *collective*.

He defines a sequence of elements $a_1, a_2, ..., a_n, ...$ as a random sequence (and calls it *collective*) if, given a property f of its elements and denoting by $n(f)$ the number of elements from the first n having the property f, the following two axioms are met:

1) When the number of terms increases ($n \to \infty$), the ratio $n(f)/n$ tends toward a limit.

2) However, if we remove, by a place selection, a part of the terms, leaving the sequence $a'_1, a'_2, ..., a'_n, ...$, the limit of the sequence $n(f)/n$ is the same, whatever the place selection is.

Axiom 2) aims to express the random character of the sequence.

It is also called irregularity axiom or the principle of impossibility of finding a playing system.

Although this definition is mathematically consistent and may pass to probability theory by introducing the relative frequency within a structure, it has also been put to conceptual criticisms.

The main criticism is that the *place selection* notion cannot be defined without coming back to the definition of *randomness,* and this weakens the consistency and the possibility of integration into an ulterior theory.

Another criticism is related to the *random* attribute that the definition proposes, asserting that this is not an absolute one because the convergence proposed by axiom 2) is not sufficient; it should include properties of the convergence speed (a sequence starting with, let us say, one million terms of 0 is not *as random as* a sequence starting with ten terms of 0 or 1).

Obviously, any definition can be improved upon, and within the community of mathematicians such attempts to represent the hazard through mathematical laws have been numerous. But all experienced conceptual flaws or were not consistent enough to reproduce the absolute randomness without question.

We will restrict our discussion to this definition sample by von Mises, underlining that, as long as a random sequence cannot be mathematically defined satisfactorily, for the term *hazard*, which is much more general and complex, the difficulty is even greater.

The fact that "ration cannot reproduce the hazard," as Borel said, still remains a principle that not even philosophers have contradicted.

We have talked here about randomness and hazard from a mathematical point of view to underline the conceptual duality of these terms.

The mathematical hazard, which we saw cannot be isolated from the philosophical hazard and cannot be consistently defined except through itself, cannot be integrated with its whole content (including the philosophical component) into probability theory.

Moreover, an event as a mathematical object is far from being a satisfactory model for the happenings from surrounding reality, so long as there is no individual definition for it to express the content of this notion. It is defined punctually, axiomatically, through the collective structure it belongs to.

As an analogy, an event is to probability theory what a point is to geometry—a simple, abstract object that is part of some set of axioms; a unit with no structural definition, which is necessary for the axiomatic constructions that create more complex geometric objects.

The model of a set of events structured as a field, although it does not reproduce even in part the absolute aleatory character of events from the real world, is the basis for the rigorous development of probability theory.

As we also stated in the section dedicated to the reconstruction of the definition of probability, this only makes sense if it is defined within a field of events. Moreover, for an isolated real event, probability makes no mathematical sense, but is a philosophical conjecture at most.

To apply probability calculus to an estimation, we must first state precisely the field that the event to be measured belongs to, and, if this is not easily observable, it must be defined or even created, if necessary.

The hazard may be philosophical or mathematical, so we deal with two probability terms: *philosophical probability* and *mathematical probability.*

This is what the terminological relativity of probability means: what we call probability of an arbitrary event in daily life, even if it expresses a quantitative character (respectively, the degree of belief), is not the same as the mathematical probability of that event (which only makes sense after we frame the event within a well-defined field).

If in a classical poker game we want to calculate the probability of one of the opponents being dealt an ace, this calculus is workable through a mathematical model of the concept (we can define the field of events as the set of parts of the set of all five-card combinations from the 52, and then use probability calculus formulas. We show how such applications work practically in the chapter titled *Beginner's Calculus Guide.*

But if we propose to estimate the probability of the event "Tomorrow it will rain with fishes," the term *event* cannot be identified with the mathematical one any more (element of a defined structure), but instead refers to a happening as an object of determinist reality, as an action of hazard.

In consequence, the afferent term of probability can no longer be identified with the mathematical concept with the same name.

In addition, the estimation will no longer be the result of a calculus, but a subjective assignation of a number at most. I

In this case, probability is the term with an exclusively philosophical content.

We may find relativities even in the mathematical concept of probability, taken individually.

Obviously, probability theory is a rigorously built mathematical theory and its notions are consistently defined.

The relativity of the probability notion comes from its definition itself.

Like any axiomatically defined concept, it is originally a product of the thinking of a mathematician, whose scientific creation,

72

although is perfectly justified in its whole by its purpose, it is finally an act of choice.

Thus, choosing one or many axioms to form the definition of a concept confers on it a relative character.

In the case of probability, exchanging the axiom of countable additivity with finite additivity, for example, creates a new type of function.

Depending on the organization and structure of the domain of this function (which also reverts to establishing a set of axioms), discrete, continuous or geometric probabilities have been defined and all these mathematical objects are expressed by the same word, probability.

There is also a relativity of the concept with respect to the representation and applicability of the mathematical model in the real world. It is about infinity.

Infinity

In the complete definition of probability (as a measure on a tribe of events), we find the countable union and the sum of a series

$(P\left(\bigcup_{i=1}^{\infty} A_i\right) = \sum_{i=1}^{\infty} P(A_i))$. Infinite sets are shown in both expressions.

Infinity also shows up in the definition of probability as a limit.

Any finite limit is attached to a convergent sequence, which is an infinite enumeration of elements.

Without the *infinite* attribute, the notion of limit makes no sense any more.

In our particular case, probability is the limit of the (infinite) sequence of the relative frequencies.

Because infinity is present in the definition of the probability notion, the question naturally arises whether the applicability of the mathematical model in the real world is justified, while this world is made up only of finite experiences.

The question also stands for all mathematical concepts whose definition is based on an infinite collective.

In mathematics, infinity axiomatically takes part in the construction of many basic notions, starting with the notion of number.

Because of its metaphysical component, mathematicians have avoided infinity for long time in their theoretical constructions.

They maximally limited its part in the process of building notions.

In our day, the infinity notion has been satisfactorily regularized in the classical domains of mathematics, but it still remains a controversial subject within set theory.

The removal of infinity from mathematics has not succeeded because, finally, it has been accepted that infinity has a logical function that is indispensable to the scientific creation itself, but is also manifest even from the level of intrinsic cognition of reality objects, in their natural, not mathematically modeled stage.

Our experience within the surrounding reality have a finite nature. Any observable object is spatially and temporally bounded.

Laboratory researchers never met infinity in their experiments.

Any experiment is performed in a finite space, within a finite time interval and with a finite quantity of matter.

With respect to micro-infinity, the natural sciences fixed the research margins a long time ago to atoms, electrons and quantums.

But a human can extend the results of his experiences at which perception takes part by using his own imagination.

The imagination, in a scientific sense, has a nearly unlimited creative power. Although imagination can only combine the results of previous experiences already stored in the memory (thus, the imagination's creative power is not unlimited in an absolute sense), it can materialize infinity at the theoretical experience level.

Let us consider the following experience:

We are observing trees at every step. We may observe smaller or bigger groups of trees and we know they are in finite number.

Even the whole number of trees we observe during our lifetime will still be finite.

We know that, if we observe a group of trees, it is possible to observe a more numerous but still finite group at another time.

On the basis of experimental observations, we study the object *tree* and may even attach a definition to it.

Such created notion of *tree* only includes our finite experience.

But if you also imagine the trees that will spring up on Earth after your own person ceases to exist (which will be observed by humans who born afterward), this theoretical experience represents an infinite number.

What we generally call experience no longer represents a limited number of determined experiences, but also represents all past and future experiences.

Thus, the notion of *tree* may include an infinite experience, which, far from denying infinity, is forced to count on it.

Therefore, infinity completes the notion of *tree* by conferring on it more intrinsic conceptual information.

The same happens in mathematics. Infinity, besides the logical function it has in the construction of notions, also has a completion function.

Let us consider a square with a 1-meter side. Its diagonal is not measurable with meter (there is no fraction of meter to fit exactly the length of the diagonal).

Then, we must be satisfied with approximations resulting from the application of Pythagoras' theorem: 1.4 m, 1.41 m, 1.412 m, ...

Which is, in fact, the length of the diagonal?

The infinite sequence of approximations 1.4, 1.41, 1.412, ... may stand for the answer.

Another answer might be the number $\sqrt{2}$, which is, in fact, a symbol representing just the previous sequence.

In this example we have infinity materialized through a length.

At other times, an infinite sequence represents an integer.

For example, the sequence of fractions 0.9, 0.99, 0.999, ... represents the integer number 1.

The affirmation regarding infinity's functions is confirmed by these simple examples: infinity is not just a simple negative finding of a lack of boundedness, but instead represents a completion operation, a coverage of an unlimited sequence of values.

Relating this function of infinity to experience inevitably generates metaphysical conjectures.

The introduction and analysis of infinity has not always led to positive results, but has led to difficulties that again raise questions on the legitimacy of using this notion in science.

The paradoxes of set theory still reveal problems today.

In the paradox of a *set of all sets*, a notion that is contradictory with itself, the completion operation is no longer valid.

The validity field of the infinity notion is still not well known.

But scientists have determined a big enough part of it to answer the current problems of science, including those of the natural sciences.

Generally, infinity has the metaphysical function of facilitating the spirit to transform the objects of experience into notions; that is, passing from a thinking level to another.

Experience put us in front of a certain number of trees and gave us the possibility of creating the notion of tree, which is not the only expression of those trees.

This notion involves the capacity of the spirit to imagine all the trees to come in the future.

This attribute of infinity provides notions with unity and value.

Infinity's function, as Kant showed in his delimitation of finite judgments, is to exclude any limitation, to give to notions the absolute fullness of their logical content by indicating not only what the notion contains, but also what it does not.

Returning to the notion of mathematical probability, we see that this concept is also defined with the help of infinity.

This notion does not raise any mathematical questions because infinity also has a well-established logical function.

The only natural question that arises is about the applicability of infinity in the practical situations of daily life, when mathematical probability is identified with the human degree of belief in the occurrence of a certain event.

As we said, all practical experiences from the surrounding reality have a finite nature.

When we want to apply the results of probability theory in daily life, whether through numerical calculations, estimations or qualitative judgments upon probabilities of various events, we must consider the compatibility of the mathematical model with the practical situation in which it is applied.

Both the mathematical model itself and its applications assume idealizations and approximations that, in turn, confer a certain relativity level to the practical results.

Conceptual and applicability relativities

Besides mathematical construction, relativity also affects the establishment of the real collective system in which the theory is applied.

The mathematical fact stated by the Law of Large Numbers, namely, that the sequence of relative frequencies converges toward probability (which stand for the definition of probability as a limit), is transposed into practice without analyzing in detail if the theorem's conditions about the infinite collective (the sequence of independent experiments is infinite) are met.

Knowing the Law of Large Numbers and the probability of rolling a certain number on a die (let us say, number 3) is 1/6, a gambler expects the die to show the number 3 once in six rolls, at least cumulatively, in a large number of rolls.

But the sequence of experiments of rolling a die in which the gambler participates is not infinite, for the hypothesis of the Law of Large Numbers to be met and its conclusion to be applied.

Even if the gambler accumulates all the experiments of this type in which he participated until that moment, he still gets a finite number of these.

Nevertheless, experience has proved statistically that the relative frequency of occurrence of number 3 on a die oscillates around 1/6 for very long successions of experiments.

If the gambler bets 5 to 1 on the number 3 (that is, he will lose the stake if he does not roll a 3 and will win five times the stake if he does), he does so because he is 1/6 convinced of the occurrence of the number 3 and is 5/6 convinced of the contrary.

These degrees of belief are, in fact, the expression of the relative frequency: in a long succession of rolls, the number 3 will occur one-sixth of the time.

This gambling behavior is amenable, with the reservation of relativity of application of mathematical model, to a regular play that generates a long succession of experiments, but it has nothing to do with a theoretical motivation in the case of an isolated bet.

And yet, most gamblers have this expectation and decision-making behavior, which is nothing more than a subjective translation of the conclusions of the Law of Large Numbers.

This example can be generalized for any type of bet for which a decision is based on relative frequencies, as well for any situation in which the subjective degree of belief is superimposed on mathematical probability: the theoretical result is transposed in a finite context (although infinity cannot be obtained at a practical experience level), conferring to the application a relative character.

Transferring the limit–probability concept in a practical situation and transforming it into a degree of belief represents at least an act of relative judgment.

Similarly, the same relativity supervenes in a case in which we consider probability as a measure and we apply its properties in practical cases of finite experiences.

The mathematical concept of measure–probability has been defined in an infinite collective, and at a practical level, infinity cannot be reproduced experimentally.

Even excluding the applicability of probability, the measure notion itself has a relative quantitative character that results from the comparative measurement (existence of a standard); the result of a measurement does not represent an absolute numerical value, but a multiple or division of the standard of measure used.

As the length of a segment measured in meters represents the number (integer or fraction) of inclusions of meter in each respective segment, the probability of an event represents the fraction corresponding to that event, as part of a sure event (with the measure 1).

That the probability of occurrence of the number 3 on a die roll is 1/6 is translated into comparative measurement terms as event *occurrence of 3* weights 1/6 from the sure event.

Still, in the classical definition of probability (the ratio between the number of favorable cases for the event to occur and the number of all equally possible cases) a relativity is involved, even from the construction of the definition.

This definition applies only to fields of events in which the elementary events are equally possible.

This attribute of *equally possible* can be textually defined as *probable is the same size* or *possible in the same measure,* which are terms that revert by their content to the term *probability* and make this notion defined through itself.

In addition, applying this definition in practical situations assumes the idealization of *the elementary events are equally possible,* which confers to probability another relativity, resulting from the nontotal equivalence of the mathematical model with the real phenomenon.

The approximation *all the die's numbers have the same probability of occurrence,* which is indispensable in probability calculus applied to more complex events related to this experiment, is questionable at a practical level and raises the question whether ignoring all physical factors equally is justified.

Assuming a die is a perfect cube, the mode of rolling can alter the aleatory attribute of the experiment. For example, repeated throws at a certain angle might favor the occurrence of a certain die's sides, as would the height from which the die is thrown, the initial impulse or the way it is held.

The unconditioned ignoring of all these physical factors comes from a certain symmetry of objects and physical actions and assumes an approximation that confers to the practical probability calculus another relativity.

Obviously, viewing the experimental results from the perspective of relative frequency in long successions of experiments (in the case of rolling a die, this operation might be performed over time by several persons, at various angles and from various heights), the relativity previously mentioned dissipates and the Law of Large Numbers finally takes effect, even if delayed.

But the *equally possible* act of approximation still remains a necessary idealization and probability theory could not be built without it.

Still related to applicability, a major relativity of probability exists that may significantly change the results of practical calculus.

It is about choosing the probability field in which the application runs.

As a mathematical object, the probability field is a triple $\{\Omega, \Sigma, P\}$, where Ω is the set of possible results attached to an experiment, Σ is a field of events on Ω, and P is a probability on Σ.

If any of these three components is changed, the result is a new mathematical object, respectively, a new probability field.

This means we can obtain different probabilities for the same event, even if it is included in different fields of events.

We take again a simple example presented earlier in this chapter as an application of the classical definition of probability.

A 52-card deck is shuffled and the first card from the upper side is face up. Let us calculate the probability for the card shown to be clubs (♣).

The experiment has 52 possible outcomes (results), from which thirteen are favorable for the event *the first card is clubs* to happen.

The probability of this event is then 13/52.

Let us now change the conditions of the experiment by assuming that we accidentally saw the last card after the deck was shuffled, and this was 5♦.

The event *the first card is clubs* now has the probability 13/51 because the number of favorable cases is still thirteen, but the number of all possible results is 51 because the last card (5♦) cannot be the first.

Therefore, for the same event E that was textually described as *the first card is clubs,* we found two different probabilities.

This double calculation is far from being a paradox, but it has a very simple mathematical explanation:

Let us symbolically denote the cards by numbers 1, 2, 3, …, 52.

In the first case (we did not see any cards), the set of all possible results of the experiment is $\Omega_1 = \{1, 2, 3, ..., 52\}$, and the probability field of events is

$$\Sigma_1 = \mathcal{P}(\Omega_1) = \{\{1\}, \{2\}, \{3\}, ..., \{52\}, ..., \{1, 2\}, \{1, 3\}, ...\}.$$

In the second case (we saw the last card), the set of all possible results of the experiment is $\Omega_2 = \{1, 2, 3, ..., 51\}$, and the field of events is

$$\Sigma_2 = \mathcal{P}(\Omega_2) = \{\{1\}, \{2\}, \{3\}, ..., \{51\}, ..., \{1, 2\}, \{1, 3\}, ...\}.$$

Σ_1 and Σ_2 do not have same elements, so they are different (for example, the element $\{52\} \in \Sigma_1$, but $\{52\} \notin \Sigma_2$).

Because the fields of events are different, the results of the corresponding probability–functions P_1 and P_2 are also different, and this explains the different numerical values $P_1(E) = 13/52$ and $P_2(E) = 13/51$.

The additional information taken into account (knowing the last card) changes the probability field and, implicitly, the probability of the event to be measured.

Although it has no apparent relationship to the example above, the estimation of the probability of an event of type *team A will win the match against team B* also submits to the same relativity of the information used.

Such probability might be estimated by taking into account one or more categories of factors such as the statistics for previous direct matches between A and B, effective teams, the playing field's physical status, current standings, and the like.

Any combination of these categories of information with the goal of estimation of probability reverts to changing the field of events and implies different final numerical results.

We can even ignore in calculus all these enumerated factors and create a rough field of events that corresponds to the following possible results: 1 (team A wins), X (draw match) and 2 (team B wins). $\Omega = \{1, X, 2\}$ and $\Sigma = \mathcal{P}(\Omega)$.

By making the approximation that these elementary events 1, X and 2 are equally possible, we get the probability
P(*team A will win the match against team B*) = 1/3.

But this result is the least relevant because such an approximation has the least possible accuracy.

Hence, same event may have different probabilities when it is included in different probability fields.

If we consider an event as completely characterized by the collective structure it belongs to, we can say without risk of confusion that in previous examples we dealt with different events, even though they are defined through same words.

We have enumerated in this section several types of relativity of the probability concept and of the application of probability theory results in daily life.

Relativity refers to the objective way in which a mathematical model is built, to the way in which this model reproduces reality and to the degree of subjectivity of estimation and interpretation of numerical results obtained from the application of probability calculus in daily life.

Although the approach to the subject of this section is scientific, with several references to the efforts of mathematics and philosophy to answer the essential problems that probability concept raises—

and the presentation is far from exhausting the subject—the goal of this excursion through the problems of relativity of probability is not didactic–theoretical, but rather is intended to catch the reader's attention and stimulate his or her thinking.

We previously aimed at outlining the image of probability as a complex but relative notion by extending the mathematical concept of probability to its relationship with surrounding reality.

The main goal is to show people who use probability results as a decision-making criterion in various circumstances that probability has no absolute character for both the mathematical concept and its transformation into a degree of belief through practical applications.

In brief, the relativities of probability follow:

1) **Conceptual relativities**
   a) *Terminological relativities*
      – Mathematical probability and philosophical probability are different objects;
   b) *Relativities of mathematical definition*
      – Defining a term through itself (the *equally possible* attribute from the classical definition);
      – The axiomatic nonstructural and nonindividual definition of event (as an element of a collective structure);
      – Choosing the set of axioms (Kolmogorov's axiomatics in the complete definition);

2) **Relativities of equivalence of mathematical model with real world**
      – The subject of philosophical probability is hazard and randomness, which cannot be mathematized;
      – Infinity, which is present in the definition of mathematical concept, is not found again in the finite experimental reality;
      – The event, as a unit of mathematical theory, does not reproduce the event from the real world, which is much more complex;

3) **Relativities of practical applications of probability calculus**

        – Choosing the field of events;

        – Idealizations of the *equally probable* type;

        – The subjective translation of the result of the Law of Large Numbers for finite successions of experiments.

These relativities require at least an additional circumspection of the person who sees probability as an absolute degree of belief and implicitly the limitation of making decisions based on the numerical value of probability as a unique criterion.

We talk more on this subject in the section titled *The psychology of probability.*

The philosophy of probability

What is the sense of the question: "What is the probability of …"?

This seems to be the essential question around which all problems of philosophy of probability revolve.

Great mathematicians like Pascal, Bernoulli, Laplace, Cornot, von Mises, Poincaré, Reichenbach, Popper, de Finetti, Carnap and Onicescu performed philosophical studies of the probability concept and dedicated to them an important part of their research, but the major questions still remain open to study:

• Can probability also be defined in other terms besides through itself?

• Can we verify that it exists, at least in principle? What sense must be assigned to this existence? Does it express anything besides a lack of knowledge?

• Can a probability be assigned to an aleatory isolated event or just to some collective structures?

These are just few of the basic questions that philosophy dealt with, through the efforts of the thinkers listed earlier, but still without a scientifically satisfactory conclusion.

Hundreds of pages of papers might be written on each of such kind of questions.

We do not pretend in this section that we review all the problems of philosophy of probability, nor do we claim that the text was optimally organized from a didactic point of view.

This presentation aims only to stimulate the research and deep knowledge tendencies of readers with regard to this subject, to complete an image of the probability concept that includes its philosophical aspects and to extend the simple image of a mathematical tool of calculus of degree of belief, which is so common among average people.

In previous sections we touched on the philosophical problems related to the construction of a mathematical model and the way in which this model reproduces the phenomenological reality.

We saw that probability may be simultaneously viewed as:

– Limit of relative frequency within a sequence of tests performed under theoretically identical conditions;
– Objective measure of possibility;
– Subjective degree of belief in the occurrence of an event.

There are also other interpretations of probability, resulting from mathematical theories with similar structures:

– Predicted relative frequency within a physical model (Drieschner);
– Measure of tendency of an experimental context to produce an outcome (Popper);
– Logical relation between a data field and a hypothesis with respect to partial implication (Keynes);
– Numerical expression of an information about the existence of an event in certain conditions (Onicescu).

All these interpretations are characterized by logical equivalences and contain elements having philosophical implications like prediction, possibility, frequency and degree of belief.

Prediction

The common enunciation, *the event* E *has probability p* generally expresses a prediction.

Through this enunciation we communicate that, in a large number of experiments, event *E* will occur in a proportion *p*.

Predictions are made about events that we do not know whether they will occur or not, at a certain moment or cumulatively, in a succession of moments.

Therefore, the subject of prediction is the unknown, in punctual or collective form.

We already talked in the section titled *Relativity of probability* about notions such as hazard and randomness and about unsuccessful attempts to mathematize them.

Comprehension of the aleatory concept is rendered harder by its ontological relativity. This relativity cannot be pushed too far before reaching affirmations like *everything is aleatory* or *there is no randomness,* although this last thesis seems more convenient.

Laplace, an adherent of absolute determinism, makes the following supposition: "In any sequence of events, no matter how irregular it is, must exist a cause from which they follow with the same necessity as the revolution motion of Earth around Sun."

Obviously, we will never be able to prove the contrary.

The belief in absolute necessity is deeply rooted in the human spirit. There will always be people who believe that fate or destiny is synonymous with necessity.

In science, such an approach would seriously render the problem of prediction more difficult because it would impose exact results.

Let us imagine an experiment involving repeated flips of a coin in identical conditions.

If the coin is of less weight, the surface on which it falls is more absorbent, and if the height from which it is thrown is lower, we can observe a preference for one or another of the coin's sides, but the individual results will remain unpredictable.

Let us admit that we could accurately measure the height from which the coin is thrown, its mass, its modulus of elasticity, and the modulus of elasticity of the surface on which it falls, and that these conditions can be maintained constant. Let us also admit that after some time we have a precise enough theoretical model to describe what happens when the coin is interacting with the surface on which it falls. Even in these conditions, we would come up with a system of equations whose rigorous solution would take a huge volume of calculations.

And this example is a piece of cake compared with the problem of predicting the side on which a die will fall or the position in which a rain water drop will touch the ground.

Briefly, the objections have a pragmatic nature: absolute determinism is not operational for science.

Randomness still has an ontological reason: it exists as a special type of disorder. Popper considers that, "For making an objective theory of probability and its application to notions like entropy or molecular disorder, it is essential to objectively characterize irregularity or aleatory disorder as a type of order. ...We may be tempted to say that hazard or disorder cannot be an objectively descriptible type of order, but they must be interpreted as a lack of our knowledge on the existing order—if such order generally exists. I think we have to resist to such temptation and we could develop a theory in which to build ideal types of regularity."

A. Cournot was the first scientist disposed to recognize an ontological reason for the hazard.

His theory about objective randomness reduces itself to:

In nature exist independent causal chains, each having its own determinism and developing in parallel, without visibly having an influence one upon each other. Sometimes they meet. Phenomena that occur at the intersection of two or more such causal chains are said to be aleatory. For example, marriage is a typically aleatory phenomenon—the two partners come from different causal chains whose intersection is something accidental.

The basic postulate of Cournot's theory is that in the universe exist independent causal successions of events: "It is a good sense principle that in nature exist successions of solitary phenomena that depend on each other and other successions that develop themselves in parallels or successively, with no dependence between them.

There are, it is true, philosophers who assert that everything in the universe is linked and they demonstrate that through subtle arguments or ingenious tricks, but they still cannot remove the common sense belief. Nobody will seriously believe that a kick in the ground will influence a ship floating at antipode or disturb

86

Jupiter's satellites. Even we would admit in theory the existence of such perturbations, these would be infinitely small, so physically they do not exist."

The effect of the causes is a damping down in time and space.

"The existence of a human being is a closed system, a small vortex with respect to the universe or other people far off in time and space."

Briefly, the universe consists of small causal systems that are included one into another. If two such systems of the same degree meet, it comes up with something aleatory.

The theory is intended to be purely realistic: the idea of hazard "is not about our mode of judgment," but "hazard exists between things."

Still, Cournot's approach raises epistemological difficulties if interpreted too consistently.

For example, an aleatory fact is not necessarily unpredictable:

"The intersection of two independent causal successions is not absolutely unpredictable. It escapes from our prediction, which covers a narrow field; but, if we would enlarge the field of prediction, we might predict some of them. But even the prediction upon them would not make them loose their fortuitous character."

Besides, it is not very clear what independent causal successions means. The *independence* notion does not seem simpler than the one of aleatory at all; moreover, it seems that the two terms could only be defined through one another.

In particular, the predictions we made in long-term independent experiments are authorized by the Law of Large Numbers, which asserts that relative frequency tends to probability.

According to Poincaré, "The facts predicted by science are only probable. No matter how rigorously we substantiate a prediction, we are never absolutely sure that experience will not deny it."

Reichenbach maintains the same point of view, although he grants the mathematical probability concept a leading part in prediction.

As long as our knowledge is probable and no natural law is directly applied to nature, but rather to a model of nature, which is idealized by neglecting an infinity of factors, the problem of knowledge cannot be solved "without admitting that applying the natural laws to reality never leads to sure sentences, but only to the probable."

Probability intervenes in knowing the nature only in "association with an ideal structure of reality."

The decision character of probability is not based on a logical construction, but on an inductive one.

According to Keynes, "The validity of an induction is based, with respect to rigorousness, not on certain information of facts, but on the existence of a probability relation. An inductive ratiocination does not assert that a certain fact will be one way or the other, but that a certain probability is assigned to it, based on a certain complex of knowledge. The validity and rational quality of the inductive generalization is then a problem of logic and not an empirical one. The effective structure of reality determines a particular form of our observing material, but it cannot decide what inductions and predictions are justified on the basis of this material."

By virtue of these ideas, probability theory, as well as induction theory, becomes a branch of logic and not a natural science, as some mathematicians and physicists are inclined to consider it.

Answering the question: "By what right do we assign probability p to an event?" reverts to answer the question: "By what right do we use the induction rule?"

An answer might be: "With no right," if we make our judgment through normal logic because this is the only justified tool we have. The induction rule (if we observe E for n times, we draw the conclusion that E will occur again at the $n+1$ time), applied to frequencies, tells us: if we observe the frequencies $f_1, f_2, ..., f_n$ being approximately equal for n times, then the results are the same for the next frequency f_{n+1}, at the $n+1$ observation.

This affirmation is based on a tacit assumption that reality is predictable, or in probability language, the limit of frequency exists.

If this is true, the induction rule will lead us to the limit. If not, this is the only tool we have for making a prediction; thus, it is relatively justified.

We do not have to resign ourselves to the logical incoherence of induction. There are always scientific methods for including an incoherent result in an existing theory, even if this unification means to extend the theory or make essential axiomatic changes.

This is also Leibniz's opinion, who says that "we should create a new kind of logics to deal with the various probability degrees."

Some logicians think they are answering these problems by taking into consideration a multivalent logic that generalizes the classical bivalent logic (in which, for a sentence p exist only two truthfulness variants— p or *non p*; the multivalent logic allows the existence of a third variant).

The accomplishment of such an attempt, even if it had guaranteed success, would be a huge act of courage because the entire logical system of science is based on bivalent logic since its creation, including mathematics, which created the probability concept.

Frequency

The results of statistical observation of the relative frequency of the occurrence of various events, which stand for the experimental base from which the mathematical model of probability started, has always fascinated scientists.

Even after the relationship between frequency and probability was mathematically established and proved by Bernoulli, through his Law of Large Numbers, the results of experience continued to intrigue by their regularity, always raising the same philosophical question: What particularly makes the frequencies stabilize in long run experiments?

Is this a law of nature and, if yes, where does its immediate motivation originate?

An experiment of repeated flips of a coin was performed 1000 times and the frequencies of occurrence of heads were recorded.

By studying the resulting statistic diagram, it was observed that the irregularity coming from a short succession of tests tends to vanish if we calculate the cumulative frequency.

If we group the 1000 observations into five successions of 200 tests each, we observe the frequency of occurrence of heads coming closer to each other than in the case of grouping into 20 successions of 50 observations each or 40 successions of 25 observations each.

Such experiments were performed and will also be performed in future. What practical statistics always noted is the fact that the ratio

$r = \dfrac{f_{max} - f_{min}}{f_{min}}$, where f_{max} and f_{min} are the maximal, respectively,

the minimal frequency of the occurrence of heads becomes lower as the volume of successions of tests grows.

In statistics, this lowering tendency of r is called stabilization of frequencies and has not been contradicted yet by experience, whether it is about tossing a coin, throwing a die or any other repeated experiment in which the initial conditions remain the same.

The fact that frequencies stabilize in experiments repeated in essentially identical conditions is empirically proved and accepted as such.

Even through the philosopher might have doubts, the experimenter is constrained to believe that, if the experiment continued forever, the relative frequencies would converge toward a certain limit.

But such conjecture could never be verified or invalidated, because we cannot perform and observe an infinity of experiments.

The experimental result has a coercive power on most thinkers: there is *something* making the frequencies stabilize.

The human brain is built so that, if we see the plot of frequencies coming close to something in a simple and regular way, we refuse to believe that this phenomenon does not have its own reason for being.

Anyone who would deny the stabilization of frequencies should try to prove that it is about that it is about appearance, that we have no guarantee of the same result in the next experiment, that is something contingent, without a rigorous explanation, a way of concealing our ignorance in academic terms.

It is theoretically possible to have no heads in 1000 coin flips.

This experimental result is not impossible because it is not contradicted by any law of physics.

Still, it is a practical certitude that this event does not happen and seems never to have happened before.

We could sustain the above affirmation even for a number of tests much lower than 1000, let us say 30, with no fear of error.

The mathematical probability of getting no heads in 30 consecutive flips of a coin is $\frac{1}{2^{30}}$, which is a number with zero followed by other 9 zeroes: 0.00000000093132...

Translated into terms of frequencies, this result sounds like this: we can expect this event (no heads in 30 flips) to happen once if we perform $10^9 = 1000000000$ successions of 30 flips each.

In a 90-year lifetime, to fulfill such an experiment would mean performing a succession of 30 flips approximately every 21 seconds, obviously with no time for anything else!

We can conclude from this that no practical experiment of such type will have that result.

A related example, which we present parenthetically as an exercise, is the following paradox that apparently runs counter to previous affirmation:

We will prove that, in theory, an experiment can be repeatedly performed by the same person so that thirty consecutive tests have the same result.

Consider the following game between two players: each player randomly picks a card from a shuffled deck and the one who gets the biggest card wins (assuming that a total hierarchy of cards is established initially).

Obviously, the probability of one player winning a game is 1/2 (as in the case of event *no heads in a coin flip)*.

We will show that a person exists who could win 30 such consecutive games ($\frac{1}{2^{30}}$ probability).

Assume we have organized a tournament for this game with 2^{30} participating players. The playing system is eliminatory (we draw for direct matches, then the first round takes place; the winners of first round must draw again for the matches of second round and so on, until the semifinals and the final; the winner of the final is declared winner of the tournament).

Such a tournament has 30 rounds and its winner plays 30 consecutive matches and wins them all.

Thus, we found the person meeting the conditions from the hypothesis.

If organizing such tournament is possible, then we could perform the succession of tests that generates the proposed statistical result.

Thus, we have built a unique succession of tests (30 games), performed by the same person (the winner of the tournament), with the pre-established result (winning the 30 games).

This would transform the event to be measured into a sure event, so the probability of 0.00000000093132 would become 1!

Beside the discussion about the ambiguous wording of the assumptions—*exist* a person who *could* win or *could* perform—which themselves might put in question the logic of the construction we made or the real possibility of putting into practice the respective experiment (we need 1073741824 persons to participate in that tournament, which means a big part of Earth's population), there is an essential judgment flaw: even if we admit that the tests are performed under identical conditions, they are not independent at all, for making a conclusion on the stability of relative frequency!

The winner plays an additional round only if he or she wins the previous round; therefore, the experiments are not independent.

Thus, this construction is not a counter-example for any statistics or law, but just a simple mathematical amusement.

The concept of the independence of tests is essential in estimating probabilities through frequencies.

In probability theory, this notion is rigorously defined: events A and B are said to be independent if $P(A \cap B) = P(A) \cdot P(B)$.

For the theory–reality interface to work practically, we must clarify the relationship between the physical notion and the probability notion of independence.

The physical independence of two events assumes there is no causal connection between them.

If two statistical collectives, one producing event A and the other producing event B, are independent, then the relative frequency of A should be approximately the same if we condition the occurrence of A by the occurrence of B.

The independence notion need not be absolute in the conditions of a practical experiment. Mathematical independence means that, in probability theory, the interactions between the components of various subsystems are not essential with respect to deducted laws.

On the same subject, Kolmogorov shows that "from a philosophical point of view, it might be more correct to talk not

92

about independence, but about the lack of importance of certain dependences in the given concrete conditions."

Returning to the experimental observation of frequencies, the belief of scientists is that something objective exists, like a natural law, which makes the curve of relative frequencies stabilize, whether the experiment is performed by one person or another, or whether now or in future.

By theoretically extending the number of experiments to infinity, it is natural to come up with the conjecture of stabilization of relative frequencies to an ideal value, which appears like a physical constant.

In this sense, probability acquires the same status as any other physical constant. Borel says that "for the physicist, probability for the radium atom to decay tomorrow is a constant as gravity acceleration is."

Possibility

An attempt at a profound discussion of the subject of *possible* cannot be considered as a venturesome act because it would stop somewhere before it starts or, in the best case, it ends in a vicious circle.

Such a discussion should reach from the beginning the category of terms *possibility–reality–existence*, which are zero-degree interdependent philosophical concepts; that is, notions that could be defined exclusively through each other and from which human knowledge cannot advance more deeply, being obliged to accept its limits.

Among all mathematical concepts, perhaps probability has the largest amount of philosophical content just because the *possibility* notion intervenes in both its interpretation and application.

In the common language, the notions *possible* and *probable* have senses attached that are different from the senses involved in a scientific theory or philosophy of probability, the frontier between them being not clearly delimited.

Sometimes, the two terms are used with the same sense. Several times we hear "It is possible, but less probable," with the implicit sense of "It is possible in principle, but I do not think it will happen." The nuanced difference is the fact that *probable* implies a more subjective and personal appreciation than *possible.*

Generally, a sentence like *p is possible* where *p* is a sentence or an event, expresses an appreciation with regard to the respective enunciation or event.

This means that the notion is a gnosiological one, by expressing certain subjective believes. But in science the concept also has a specific ontological ground.

Let us consider the enunciation: "It is possible by rolling this die to obtain the side with number 4."

Can we attach a truth value to this sentence?

We may do that, in an experimental way: we roll the die; if it shows 4, we are right; if it does not, no.

But through the initial enunciation we did not state that we would obtain a 4 at *this* roll. If we had intended this, we would have said, "Now a 4 will come."

We stated that it is possible to roll a 4, meaning that we refer to an arbitrary roll. In this case, the possibility affirmation has an existential character. It is logically equivalent with "A 4 will come at roll number 1, *or* at roll number 2, *or* at roll number 3, etc."

Now the sentence may be true or false in a theoretically logical sense. That it could be true is obvious: it is enough to roll a single 4 in one of the rolls. That it could be false is much harder to understand.

The possibility affirmations are empirically verifiable, but they are not falsifiable.

Indeed, to deny the initial enunciation, we should affirm: *It is impossible to roll a 4.* Because it is clear that we do not deal here with a theoretical impossibility, thus: *It is impossible* can only be interpreted as *It is physically necessary a 4 not to come.*

Thus, we observe that it is about a physical possibility.

Any affirmation of that type ("It is possible that…") contains an existential assertion and implicitly admits an experimental context and an alternative.

In the example considered, the alternative might be (1, 2, 3, 4, 5, 6) die's sides or any subalternative derived, for example (4, non-4).

The experimental context is the fact that the die does not disintegrate after a roll and can be used many times.

94

A sentence like: "It is possible by rolling this die to obtain the side with number 4" can only be thought of as a sequence, a family of sentences of the type: "At roll number *n*, 4 occurs." Even if 4 has not occurred by now, the possibility continues to exist.

The scientist is constrained to think to of all possibilities as virtually existing, even each time only one occurs. In such acceptation, possibility is only a modality of the future.

Any appreciation of the physical possibility of an event is related to prediction.

From empirical experience or from a theoretical model, we know that we have no reason to exclude the event from our view and that it will finally occur in a long-running succession of tests.

The logical mechanism leading to the assertion of the type: "It is possible 4 to come" is something like: In certain conditions, 1, 2, 3, 4 ,5 or 6 occurs.

We do not know the initial status of the die, nor how it will roll on the table. Therefore, we must exclude none of the 6 cases. Thus, rolling a 4 is possible.

When planning an experiment, the scientist foresees the alternative, consciously or not, in a more or less precise manner.

Its estimation, that is, possibilities, is a first step in planning the experiment. For the scientist, they exist in reality and in this sense the *possible;* namely, the alternative, has an existential reason.

Without this ontological supposition, science would not be possible.

Many times, appreciating something as possible is grounded on observation of some necessary but not sufficient conditions for that something to occur. This rule has a methodological value in analogy reasoning.

The mathematical conjectures and theorems start just from the intuition *p is possible,* which is generated by observing some necessary conditions for *p* to be true.

Fortunately, in mathematics things do not stop there: the sufficiency of the conditions discovered must also be proved and, if not, the conjecture must be invalidated through a counter-example.

The judgment *event* E *is possible* is exclusively qualitative, and every qualitative judgment is only a beginning of knowledge.

The scientist also wants to give a sense to the question: "How possible is *E?*" The measure of the possibility of *E* is its probability.

Most thinkers agree that *probable* is a measure of *possible,* but they do not agree on the interpretation of these words. How such a measure can be estimated? Does the *real probability* of an event exist? Does it represent more than a simple opinion?

Most of the top probability scientists have put these questions to themselves for more than a century.

The interpretation nuances of the two terms *possible* and *probable* might generate a classification that is a kind of analogue to *abstract, theoretical possibility* and *concrete possibility* (in which exist conditions that make us expect an event to happen).

On other hand, *probable* is also felt like something opposite to *sure* and is kind of synonymous with *credible* or *verisimilar.*

It follows that in daily life, the main sense of the word is subjective, of personal expectation, of partial belief that something will happen or has already happened.

Even an etymological analysis of the word *probability* shows that it originates in the Latin verb *probo, probare,* which means *to verify.*

Latin also uses the word *verisimilitudo* for *probability* and *verisimilis* for *probable,* and this last one was assimilated by other languages with the same sense, in English, *verisimilar.*

In probability theory, *verisimilitude* has a well-stated sense: the probability of an event coming from a hypothesis, the so-called *conditional probability.*

What does verisimilar actually mean in reality? From the history of analysis of this concept, and further from antiquity we have the famous example of Carneade:

A boy goes into a dark cellar. He sees there a rope that looks like a coiled snake. The boy gets scared, not knowing whether the rope is a snake or not. He waits a little and observes that the object does not move. Maybe it is not a snake. He carefully advances a few steps and the object still remains motionless. His belief that it is a lifeless object increases. He takes a stick and touches it. The object does not move. Only when he takes it in his hand does he reach the practical certitude.

This example contains in itself the subjective probability promoted by probability scientists such as de Finetti. The example shows very intuitively the evolution from uncertainty to practical certainty via Bayes's theorem.

Carneade's rivals raised the following question to him:

Verisimilitude is something resembling the truth. If you do not know what the truth is, how can you know that something looks like it?

These polemical requests made Carneade develop a nuanced analysis of the concept of verisimilitude. He considered that the degree of belief attached to things is given by:

– Vivacity of sensations
– Order of representations
– Absence of internal contradictions.

In the rope example, at the beginning the vivacity of sensations was not enough (it was dark there), then the boy appealed to the order of representations by verifying whether the rope was moving, and the certitude came as result of finding no internal contradictions between rope's representation and the object in front of him.

If a mouse under the rope had made it move, the boy's level of belief in the hypothesis *it is a rope* would have diminished owing to the internal contradictions between representations: a rope does not move by itself.

Carneade also says: "We, humans, cannot reach the absolute truth through reason, nor perception, as skeptics proved. The truth, the certitude are not given to us. Still, life requires us to act. Therefore, we have to be content with the practical appearance of truth—the verisimilitude."

As we mentioned from the beginning, this presentation aims to be a short journey through the problems of the philosophy of probability, not with the goal of exhausting the subject—which as a practical matter is not possible—but to stimulate the reader's conceptual thinking and capacity for profound study.

Although this guide is didactically focused on the application of probability theory to practical situations of calculus, we considered the creation of an objective image of probability necessary.

And an objective enough image should show even the subjective aspects; even these belong to its philosophy.

The philosophical problems raise no doubts about probability theory with respect to its mathematical construction, but they put in question the idealization elements of the notions involved, especially the way these notions represent the surrounding reality.

In any knowledge act, the reflection of an objective fact is involved, as well as certain biological structures specific to humans.

This is just why knowledge is exclusively human.

As concerning the interpretation of probability concept, this is considered to be something objective.

The uncertainties in interpretation come from the fact that theory is applied to two distinct types of problems: the *possibility* problems, which have an objective value, and the *probability* problems, which have a subjective nuance.

Probability has a double meaning: first as a measure of the real possibility of things (the physical probability revealed through frequency) and second as the degree of trust; in other words, there exist a philosophical probability and a mathematical one, and these are not to be confounded.

The probability of an event does not really exist in the phenomenal world, like mass, force and the Greenwich meridian do not exist as real objects. It only exists abstractly.

Its objective significance is that, starting from the same hypotheses, all mathematicians will find the same value for it, no matter the individual subjective opinions.

It serves as a tool for acquiring a partial knowledge of the surrounding world, which is not equivalently and totally reproduced, simply because the hazard cannot be theoretically modeled and quantified.

And then what is the justification for probability theory? What is the sense of its application?

Humans are sentenced to act in uncertainty conditions. If humans had an infinite intelligence and calculus capacity, he or she could predict the future and would know our entire past.

Probability theory is the mathematics of idealized hazard.

Its application consists of reducing all events of a certain type to an arbitrary number of equally possible cases and calculating the number of favorable cases.

Probability is nothing more than the mathematical degree of certainty we have about an event.

It is simultaneously objective and subjective.

Probability does not exist beyond us.

In fact, it is not about the degree of certainty we have a priori, but one we should have if we were perfectly rational and could make the *equally possible* judgment.

Therefore, probability is the only reasonable way to behave in conditions of partial knowledge and uncertainty, by using mathematics as a unique method, which is rigorous and unanimously accepted.

The Psychology of Probability

From the previous sections dedicated to relativity and philosophy of probability, we realize that this concept has a major psychological component generated by its impact with the human mind at the cognitive level.

The probability notion itself, through its interpretation as a degree of belief, as well as the application of probability theory in daily life, are the subject of objective or subjective human appreciations and judgments, no matter the human's level of knowledge.

The human mind is built so that it manifests through two apparently contradictory tendencies: on the one hand, it is eager for knowledge and disposed to the mental effort of searching for answers to questions about phenomena from the surrounding world.

On the other hand, it accepts the comfort of immediate explanations and theories that do not contradict other convictions, at least at first.

In this sense, the interpretation of probability as an absolute or at least a sufficient degree of belief (to make decisions) has a partially solid motivation: thus far, probability theory is the only valid theory operating upon aleatory events (even if in idealized context) through incontestable mathematical methods.

This motivation is in fact the expression of the comfort we just talked about: humans search around them for explanations, causes and theories to answer to their questions conveniently and chooses the most rigorous (but not always).

They limit their mental effort to this plane search (on the horizontal), by omitting (intentionally or not) that another dimension of cognition (on the vertical) exists where profound study is done through abstraction and generalization.

Obviously, this last type of cognition process assumes judgments that are not at everybody's hand, a certain level of education, intelligence and conceptual perception being required.

Another factor that influences the thinking process is the subconscious, this copilot of the human mind that much of the time, without our being aware, takes control over the functions of the organism, including the cognitive ones.

The two distinct notions—philosophical probability and mathematical probability—even when individually perceived, are frequently confused by a person in situations involving theoretical judgment or applications.

Besides this major conceptual inconsistency, qualitative interpretation errors may be multiple:

- The exclusive use of the term *probability* in the sense of its classical definition;

Not every field of events can be reduced to a finite field with equally possible elementary events, for the classical definition to apply.

- Attaching a probability to an isolated event;

The field of events as a domain of the probability function must be structured as a Boole algebra. The probability of an event does not make sense if that event does not belong to such a field.

- Isolation of probability as a function from the probability field;

Probability is determined by the triple (the set of possible outcomes–the field of events–the probability function); namely, the probability field. Defining the probability of an event means putting in evidence each of the three components, not just the function.

- Identification of probability with the relative frequency;

Although probability is the limit of relative frequency and prediction can be made only for a long-running succession of experiments, the result is frequently applied by analogy to an arbitrary succession of experiments and even to an isolated event.

- Transformation of mathematical probability into an absolute degree of belief;

Neglecting the relative aspects of probability turns it into an absolute criterion in making decisions in various situations that require an action.

These interpretation errors are related to the intellectual capacity as well as to the intimate psychological mechanisms of the subject, including the functions of the subconscious.

Among them, the most important interpretation error from a psychological point of view is the transformation of probability into an absolute degree of belief, which also has social implications, because it results immediately in the acceptance of probability as a unique decision-making criterion with effects in the sphere of personal actions that also affect other people around the subject.

Including this psychical behavior in the category of qualitative interpretation error finds its complete motivation in the section titled *Relativity of probability*, where we saw that the probability notion has a multirelative character, with respect to both the concept itself and to its equivalence relation with the phenomenal world it models.

Making an optimal recommendation about this line of decision-making conduct is a complex problem, which itself represents another individual theory.

It is easy to make a recommendation at an immediate classification level: changing the attribute of *absolute* into *relative*.

The subject must mentally perceive the (mathematical) probability as a rigorously calculated degree of belief, but *relative* with respect to the possibility of the physical occurrence of an event.

This attribute assumes in succession the acceptance of probability as a decisional criterion, but not exclusively. The decision come from a complex of criteria, perhaps weighted, some of which may be even subjective.

Mathematics is still firmly involved in sustaining such recommendation and also in an eventual theory of subjective probability.

De Finetti declared himself with no doubt as sustaining a subjectivist concept of probability: "My point of view may be considered as the extreme of subjectivistic solutions. The purpose is to show how the logical laws of probability theory can be rigorously established from a subjective point of view; among other things, it will be shown how, although I refuse to admit the existence of any objective value and signification of probabilities, we can make an idea about the subjective reasons due to which, in a big part of different problems, the subjective judgments of normal people are not only not much different, but they rigorously coincide with each other."

He obsessively repeats in his papers that "Probability does not exist," being completed by Barlow with "...except in our mind."

De Finetti proposes an extremely simple definition for probability: "Assume a person is constrained to evaluate the ratio p at which he or she would be disposed to change an amount S, which may be positive or negative, depending on the occurrence of an event E, with the sure possession of amount pS. Then, p is said to be the probability of E given by the respective person."

This manner of defining probability is, obviously, disputable.

First, the definition makes no sense if referring to an empirical subject, for the simple reason that ratio p does not actually depend only on event E, but also on the amount S, as psychological research has revealed: if a normal person feels indifferent about the alternative *$1 for sure against $6 if a 6 occurs at a die roll*, an alternative of the type *$10000 for sure against $60000 if a 6 occurs at a die roll* would not be felt so indifferently by the same person, who would prefer the first variant.

But just the hypothesis p = p(E) and not p = p(E, S) is essential for de Finetti's axiomatization (p is a function of E and not also of S).

This can be also observed in a demonstration of the additivity of p, which is rigorous only if p does not depend on S.

At this objection, the author answers: "It would have been better if I would deal with utilities, but I was aware of the difficulties of bets and I preferred to avoid them by considering small enough stakes. Another lack of my definition—or, better stated, of the tool I choose to make it operational—is the possibility for those accepting the bet against the respective person to have better information. This would lead us in game theory situations. Of course, any tool is imperfect and we have to be content with an idealization ...

Probability theory is not an attempt to describe the real behavior of people, but refers to coherent behavior and the fact that people are more or less coherent is not essential."

But let us admit that we can skip the drawback of initial definition by postulating that this is the way an ideal person would act (the ratio p to only depend on E), a person who is not interested in winning, but uses the bet only to clarify his or her own initial subjective probabilities that persist in the subconscious.

Once this deadlock is passed, de Finetti develops an elegant and coherent theory about which he says that it succeeded in formalizing the probability concept that is the closest to the one used by common people, the one used by people in their practical judgments.

The found rules "are in fact only the precise expression of the rules of logics of probable, which are unconsciously, qualitatively, even numerically applied by all people, in any life situation."

As a conjecture of conclusions on his theory, de Finetti puts the following question: "Among the infinity of possible estimations, is there one we could consider as objectively coherent, in an undefined sense for the moment? Or, can we say at least about an evaluation that is better than another? ... For assigning an objective sense to probability notion, during the centuries two schemes were imagined: the scheme of equally possible cases and the frequency considerations. But none of these procedures obliges us to admit the existence of an objective probability. On the contrary, if someone wants to force their significance for reaching such conclusion, will face the well-known difficulties, which are vanishing themselves if we become less pretentious, that is if we will try not to eliminate, but to define the subjective element existing between them ... The problem consists in considering the coincidence of opinions as a psychological fact; the reasons of that fact can retain their subjective character, which cannot be left aside without raising a lot of problems whose sense is not even clear."

Regarding the decisional criteria based on a degree of belief, it has been ascertained that a human resorts psychologically to practical statistics in large measure.

As in the case of using mathematical probability as an absolute decision-making criterion, consulting the statistical results often becomes the unique criterion for making a decision in certain situations.

For example, a person may decide to undergo a certain operation if the medical statistics reveal a satisfactory success rate (let us say, 80 percent) for this type of operation.

Such a decision is based exclusively on the results of previous practical statistics, which the undecided person deems trustworthy.

Many times decisions are also made on the basis of previous unrecorded statistics whose results are established through personal observations and intuition.

For example, when we observe black clouds in the sky, we hurry to get home. The statistics used here as the basis of a decision are the set of personal observations performed during a lifetime of the same phenomenon: in most situations (p% from them, p enough big), it rained when such clouds were in the sky. Then, it is very possible (p% degree of belief) that it will rain again now, so we quicken our steps to get home sooner.

Such decisions are not, in practice, the immediate result of consulting the statistics, but are a process that transforms the statistical result into a degree of belief, which is still an expression of a probability.

Probability has a direct theoretical connection with statistics (mathematical statistics is an extension of probability theory and is even considered a part of it), but is also connected to practical statistics.

Practical statistics means a collection of outcomes (results) of a certain type of experiment, recorded over time.

These results correspond to a finite time period (that explains the use of the term *statistics until this moment)*, and therefore to a finite number of experiments.

Generally, solid practical statistics cover a very large time period and, by implication, a very large number of experiments.

The statistical results are recorded in diagrams or bidimensional tables, in which time (experiment's number) is one of coordinates.

Such representation shows the relative frequency of the event or events that are the subject of statistics, even though this frequency is not directly calculated (as a ratio), with the condition that the experiments are performed in identical conditions.

From here, the connection with probability is immediate. Being a relative frequency, the more results the statistical record contains, the better this frequency approximates the probability of a respective event.

Thus, the degree of belief of the person making the decision based on statistics becomes a translation of a partial result (the relative frequency recorded by statistics until that moment) of an isolated event (respective decision-making situation).

The subject, in fact enlarges the succession of statistically recorded experiments with another one, the one he or she is involved in that has not yet happened, by applying the relative frequency corresponding to the previous n experiments to the $n+1$ experiments including the virtual $n+1$ experiment.

This in inductive reasoning resembling prediction based on frequential probability.

In the previous medical example, the patient makes the following judgment: If the operation succeeded in 80 percent of previous cases, it will also succeed in my case with 80 percent certainty, so I will have the operation and assume the 20 percent risk.

In reality, this risk might be lower or higher because of other criteria not taken into account. The 80 percent certainty does not come from objective factors specific to the isolated medical case of that patient, but from practical statistics, which correspond to other persons.

Assuming that the respective operation is not influenced by the medical condition of the patients, the operation is performed identically from a technical point of view for all patients and all surgeons are equally well trained (idealizations that are necessary to ensure the identical conditions in which the tests are performed), we can affirm that the 80 percent statistical result stands for the relative frequency of the event *operation succeeds* for the n tests.

If the medical record contains a large enough number of cases (n is large enough), we can state that the respective relative frequency approximates the mathematical probability of the event studied.

In these ideal conditions, the transformation of *probability* resulting from statistics into a degree of belief in the occurrence of the predicted event is not incorrect, having the same status as a probability, but with the same reservations this kind of interpretation implies (we already agreed that the *absolute degree of belief* attribute is contradictory to the multirelative character of the probability notion, whether mathematical or philosophical).

This analogy with the inductive probability prediction is still not perfect. *Probability* that is generated by the relative frequency of a practical statistic is in fact a fixed number and depends on the number of recorded tests (it is, in fact, the last calculated relative frequency), while probability as a limit of relative frequencies assumes the existence of such a sequence, at least theoretically.

Many times, a practical statistic is chosen as a decision-making criterion instead of mathematical probability, even when the latter can be rigorously calculated.

This decision-making behavior belongs especially to gamblers.

If a person who regularly plays a certain game of chance makes his or her own statistics (or has access to other statistics to consult) about a certain gaming event, that person tends to transform the statistical result (the relative frequency of occurrence of an expected event) into a degree of belief that influences his or her decision, instead of mathematical probability, which can be calculated.

It is well known that, in gambling, most of probabilities attached to gaming events can be calculated, and gamblers may consult guides that contain collections of such results (like the chapter "Applications and results for the games of chance").

Still, most of them prefer to choose practical statistics as a criterion for their gaming decisions.

Here is a hypothetical example:

A regular Texas Hold'em player has developed his or her own statistics from various experiences in gaming situations.

In one game, he is dealt an ace. Before the flop, he must make a decision about his bet (raise or call). This decision is influenced by his degree of belief in the occurrence of the event *the flop cards contain at least one ace.* By consulting his own statistics, he finds a relative frequency of 29 percent for that event in each respective gaming situation and he takes this result as degree of belief without resorting to other information. On the basis of these figures, he decides to raise.

If the player had resorted to the real mathematical probability for the respective event, which is 18.27 percent (see the section titled *Texas Hold'em Poker* in the chapter on applications for the games of chance), his decision might have been different.

Of course, if the player's own statistics contained a very large number of records of respective gaming situations, the figures would not be so different from the real probability, owing to the principle of stabilization of relative frequency in long-running tests.

Appealing to practical statistics with the goal of making decisions is a specific characteristic of the human mind and is plenarily manifest in daily life.

Even if we do not always observe this thing, people around us consistently make more decisions based on statistics rather than on probability.

Here is another simple example from daily life:

A policeman making his guard sees a masked man with a bag on his back running out of a store. The policeman makes the decision to run after that man and arrest him.

Obviously, the policeman's belief that results in his action is the fact that that man is a delinquent.

This belief is not an absolute one, but it hides a degree of certainty that also comes from a call to statistics.

The policeman ascertains that, in a very large majority of cases, the man proved to be a delinquent; in other words, the degree of belief acquired as result of consulting statistics is very high.

The policeman does not take into account any other alternatives—that person might be the store's owner who has to go to a masked ball and is in hurry—but he immediately makes his decision on the basis of the degree of belief provided by statistics.

What is, in fact, the motivation behind such general behavior of appealing to statistics? Is there a rational motivation or it is only a matter of human biological structures?

The answer is somewhere at the middle and can be explained in large part by psychology.

All along, man felt safe as an individual only when he was grounded on something sure and perceptible. The human mind also submits to this principle.

Practical statistics is a collection of *sure* results; namely, frequencies of events that *already happened*, and these happenings are a certainty.

Unlike statistics, the prediction, by estimating a degree of belief, refers to events that have *not happened* yet and their occurrence is an uncertainty, so the human mind classifies them in the category of *unsafe* and tends to increase somehow their sureness by transferring sure things (the statistical results) upon them.

Although humans act in the real word in uncertain conditions, in an unsure environment, the human mind perceives this thing as an anomaly and tries to ameliorate it by elaborating degrees of belief that come from sure environments, such as practical statistics.

This psychic process of unconditioned migration to certainty environments is a reflex action of possibly ancestral origin and may be changed only through a profound study of notions of frequency, probability and degree of belief.

Beyond the psychological aspect, there is also a minimal rational theoretical motivation behind this process of appealing to statistics, which results from its connection to probability and is analogue to the motivation for probability-based decisions in games. We talk about this more in the chapter titled *The Probability–based Strategy.*

Appealing to practical statistics is not only an individual option, but it is also currently practiced in industry and even in scientific communities.

Economic fields such as insurance, marketing, industrial testing of materials; research domains such as pharmacology and medicine; and disciplines such as psychology and sociology all use statistics as the main tool in making decisions related to their activities by transforming statistical results into degrees of belief and decisional criteria.

Obviously, as the statistically recorded segment grows over time, the new results may change the degree of belief in the occurrence of expected events.

Returning to the example of the patient who consults the medical statistics in order to make a decision about an operation, no matter what the rate of success of that operation until that moment, a new statistical record (a new surgery case) will change the patient's degree of belief in the success of his or her operation, whether the new case was a success or not.

In other words, the patient will estimate a probability of success (as degree of belief) other than the one before the last statistical record.

Although probability as a degree of belief is something subjective at the psychological level, as revealed by the previous example, the phenomenon described has a rigorous mathematical model, namely, Bayes's theorem.

Bayes's theorem is a main result in probability theory, which relates the conditional and marginal probability of two aleatory events A and B.

In some interpretations of probability, Bayes's theorem explains how to update or revise beliefs in light of new evidence.

The probability of an event A conditional on another event B is generally different from the probability of B conditional on A (see the section titled *Independent events. Conditional probability* in the mathematical chapter).

However, there is a definite relation between the two, and Bayes's theorem in its simplest form is the statement of that relation.

As a formal theorem (that is, a statement of a formal system, namely, a system that captures the essential features of a real-world object or phenomenon in formal language, which is a set of finite-length words drawn from some finite alphabet), Bayes's theorem is valid in all interpretations of probability.

However, frequentist and Bayesian interpretations disagree about the kinds of variables for which the theorem holds.

So, if event B is expected (with a certain given probability) as result of a test and event A occurs instead, the question Bayes asked was: How does the probability of B change if we know whether or not A has occurred? Or: How does the observation of whether or not A has occurred cause you to update your probabilistic assessment of the likelihood of B occurring?

In short, we want a helpful expression for $P(B|A)$.

We already have the definition of this expression:

$$P(B|A) = \frac{P(A \cap B)}{P(A)}.$$

Considering the analogue expression for $P(A|B)$ and reducing the common term we find $P(B|A) = \dfrac{P(A|B) \cdot P(B)}{P(A)}$.

This is Bayes's theorem. It is nothing more than the definition of conditional probability applied a couple of times plus a little bit of algebraic cleverness.

The importance of Bayes's theorem is that the informational requirements to calculate $P(B|A)$ from Bayes's theorem are different from those required by the definition of $P(B|A)$.

Each term in Bayes's theorem has a conventional name:

$P(A|B)$ is the *likelihood* of A given B for a fixed event B, respectively the conditional probability of A given B.

P(B) is the *prior probability* or *marginal probability* of B. It is prior in the sense that it does not take into account any information about A.

$P(B|A)$ is the conditional probability of B, given A. It is also called the *posterior probability* because it is derived from or depends upon the specified event A.

P(A) is the prior or marginal probability of A, and acts as a *normalizing constant*.

With this terminology, the theorem may be paraphrased as

posterior = (likelihood x prior) / normalizing constant

In other words: the posterior probability is proportional to the prior probability times the likelihood.

There is also another form of Bayes's theorem, using complementary events:

$$P(B|A) = \frac{P(A|B) \cdot P(B)}{P(A|B) \cdot P(B) + P(A|B^{c}) \cdot P(B^{c})}$$

More generally, if $(A_i)_{i \in I}$ is a complete system of events (a partition of the field of events), then:

$$P(A_i|A) = \frac{P(A_i) \cdot P(A|A_i)}{\sum_{i \in I} P(A_i) \cdot P(A|A_i)} \text{ , for every } i \in I.$$

These are the simplest forms of Bayes's theorem, which is a simple formula itself.

Despite its simplicity, it stands for a rigorous model of the relativity of probability as a degree of belief and reflects how this probability changes when taking into account real statistical information.

This theorem has several practical applications. One example of how Bayes's theorem can be used is:

All people from a small community are tested for a medical condition (HIV, for example).

Suppose you have a laboratory test for this condition and the following information:

– The chance of a random draw from the community having the medical condition is 1/1000.

– The chance of a false positive test result from the lab test is 1/100.

– The chance of a false negative test result for the lab test is 1/500.

Without the test, any person rationally believes his or her chances of having the medical condition are 1/1000.

An important question to someone just tested is: If the test comes back positive, what are the chances of having the medical condition?

To answer this question, we will formalize the information provided above. Define the following events:

A : The person has the medical condition.

A^C : The person does not have the medical condition.

B: The person has a positive test result.

B^C : The person has a negative test result.

Since 1 out of every 1000 people have the medical condition, $P(A) = 1/1000$.

Since there are false positive results in 1 out of every 100 lab tests, $P(B|A^C) = 1/100$.

And, since there are false negative results in 1 out of every 500 lab tests, $P(B^C|A) = 1/500$.

But we can go further. The data provided also allow us to calculate:

Since 999 out of every 1000 people do not have the medical condition, $P(A^C) = 999/1000$.

Since there are 99 correctly positive lab tests out of every 100 positive lab tests, $P(B^C|A^C) = 99/100$.

And, since there are 499 correctly negative lab tests out of every 500 negative lab tests, $P(B|A) = 499/500$.

Remember that the person just receiving the results of the lab test is interested in calculating the probability of having the medical condition, having just heard a positive lab test result.

In terms of our notation above, the person wants to calculate $P(A|B)$.

Bayes's theorem tells us,

$$P(A|B) = \frac{P(B|A) \cdot P(A)}{P(B|A) \cdot P(A) + P(B|A^C) \cdot P(A^C)}$$

By replacing our known values, we find $P(A|B) = 0.0908 = 9.08\%$.

This result says that, having just tested positive for the medical condition, there is only a 9.08 percent chance that the person actually has the condition.

Is there some intuition behind this result? Suppose there are 1000000 people randomly chosen from the population, all of whom are tested for the medical condition.

We would expect 1000 of them to have the medical condition.

Of the 1000 who have the condition, two will not show a positive test result. This is what a false negative lab test means.

But we also know that 998 of the people who have the medical condition will correctly get a positive test result.

Of the 999000 who do not have the condition, 9990 will receive a positive test result. This is what a false positive means.

So, a total of $998 + 9980 = 10978$ receive a positive test result.

But of this group only 998 are actually positive.

So, the probability of having the condition if the test is positive is 998/10098, or 0.0988. Pretty close.

Although we said that Bayes's theorem is a rigorous model of how the degree of belief changes with information (and this is a certitude), in fact it reveals the relativity of probability through psychological perception.

It does not create a unique probability for the conditioned or unconditioned event, but creates another probability function on the same field of events, thus creating a new probability field:

If $\{\Omega, \Sigma, P\}$ is the prior probability field and $A, B \in \Sigma$ with $P(B) \neq 0$, the probability of event A conditional on event B is

$$P(A|B) = \frac{P(A \cap B)}{P(B)}.$$

If we fix the variable B we obtain a function $P_B : \Sigma \to [0,1]$, $P_B(A) = P(A|B)$ (called the likelihood function).

It can be proved that the triplet $\{\Omega, \Sigma, P_B\}$ is a probability field.

Bayes's theorem only provides information for the expression of function P_B in this new probability field.

It does not provide any absolute value of the prior probability $P(A)$, nor of the posterior probability $P(A|B)$, and this lack reverts again to the relativity of probability as degree of belief.

The use of Bayes's theorem in practical situations is another example of appealing to statistics to achieve a degree of belief, but this time in a mathematical manner.

While humans appeal to past statistical results to get a subjective probability as a degree of belief, there is also a reverse psychological process: the prediction of future statistical results based on a given probability.

Such prediction behavior manifests preponderantly in gambling, where probabilities are associated with stakes in order to predict an average future gain or loss.

This predicted future gain or loss is called *expectation* or expected value and is the sum of the probability of each possible outcome of the experiment multiplied by its payoff (value).

Thus, it represents the average amount one expects to win per bet if bets with identical odds are repeated many times.

A game or situation in which the expected value for the player is zero (no net gain nor loss) is called a *fair game.*

The attribute *fair* refers not to the technical process of the game, but to the chance balance house (bank)–player.

For example, an American roulette wheel has 38 equally possible outcomes. Assume a bet placed on a single number pays 40–to–1 (this means that you are paid 40 times your bet and your bet is returned, so you get back 41 times your bet). So, the expected value of the profit resulting from a one dollar repeated bet on a single number is:

$$\left(-\$1 \times \frac{37}{38}\right) + \left(\$41 \times \frac{1}{38}\right), \text{ which is about } -\$0.10526.$$

Therefore one expects, on average, to lose over ten cents for every dollar bet.

For a one dollar bet on heads at a coin flip (win $1 and your bet is returned if heads comes up, and loose the bet if not), the expected value is: $\left(-\$1 \times \frac{1}{2}\right) + \left(\$1 \times \frac{1}{2}\right) = 0$, so this is a fair game.

This prediction also has a rigorous model in probability theory, namely, the mathematical expectation:

Let X be a discrete random variable (see the definition of a discrete random variable in the dedicated section in the mathematical chapter) with values x_i and corresponding probabilities p_i, $i \in I$.

The sum or sum of series (if convergent) $M(X) = \sum_{i \in I} x_i \cdot p_i$ is called mathematical expectation, expected value or mean of variable X.

So, the mathematical expectation is a weighted mean, in the sense of the definition given above.

The psychological factor manifests not only in the interpretation and application processes of probability, but also in the estimation of numerical results.

The calculation of mathematical probability itself in practical situations is the subject of subjective and even erroneous judgments in turn, especially if applied without minimal conceptual knowledge of probability theory.

Many times probabilities are intuited, even when a rigorous calculus is possible. In such situations, the respective person estimates or compares two probabilities without executing the entire process, which consists of framing the problem, studying it and applying a rigorous calculus, but bases the decision only on his or her own feeling.

Many times, sufficiently good approximations are obtained through intuition, but most of time the result does not conform to reality and may lead to wrong decisions.

The estimation of probabilities may be incorrect at the intuitive level because such an intuition is not in fact a quick rational process that skips some steps and reaches a correct final result, but is a group of reflex actions that combine known trivial results with a human's own subjective expectations.

The estimation errors can come from an unsubstantial observation, an inadequate or inexistent framing of the problem or they may be just technical errors in calculus.

Here is a simple example of false intuition, in which the error comes from an incorrect framing of the problem:

You have the following information: A person has two children, at least one a boy. What is the probability of the other child being a boy, too?

You are tempted to answer 1/2, by thinking there are only two choices: boy and girl.

In fact, the probability is 1/3, because the possible situations are three:

boy–girl (BG)

girl–boy (GB)

boy–boy (BB)

and one of them is favorable, namely (BB).

The initial information refers to both children as a group and not to a particular child from the group.

In the case of estimation of 1/2 , the error comes from establishing the sample space (and, implicitly the field of events) as being {B, G}, when in, fact it, is a set of ordered pairs: {BG, GB, BB}.

The probability would be 1/2 if one of the two children had been fixed by hypothesis (as example: *the oldest is a boy* or *the tallest is a boy*).

For many people, the famous birthday problem is another example of contradiction with their own intuitions:

If you randomly choose twenty-four persons, what do you think of the probability of two or more of them having the same birthday (this means the same month and the same day of the year)?

Even if you cannot mentally estimate a figure, intuitively you feel that it is very low (if you do not know the real figure in advance).

Still, the probability is 27/50, which is a little bit higher than 50 percent!

A simple method of calculus to use here is the step by step one:

The probability for the birthday of two arbitrary persons not to be the same is 364/365 (because we have one single chance from 365 for the birthday of the first person to match the birthday of the second).

The probability for the birthday of a third person to be different from those of the other two is 363/365; for the birthday of a fourth person is 362/365, and so on, until we get to the last person, the 24th, with a 342/365 probability.

We have obtained twenty-three fractions, which all must be multiplied to get the probability of all twenty-four birthdays to be different. The product is a fraction that remains as 23/50 after reduction.

The probability we are looking for is the probability of the contrary event, and this is 1 − 23/50 = 27/50.

This calculus does not take February 29 into account, or that birthdays have a tendency to concentrate higher in certain months rather than in others. The first circumstance diminishes the probability, while the second increases it.

If you bet on the coincidence of birthdays of twenty-four persons, on average you would loose twenty-three and win twenty-seven of each fifty bets over time.

Of course, the more persons considered, the higher the probability.

With over sixty persons, probability gets very close to certitude.

For 100 persons, the chance of a bet on a coincidence is about 3000000 : 1.

Obviously, absolute certitude can be achieved only with 366 persons or more.

One of the most curious behaviors based on false intuition is that of lottery players, where the winning probabilities are extremely low.

As a game of chance that offers the lowest winning odds, it is not predisposed to strategies. The player (regular or not) purely and simply tries his or her fortune, whether he or she knows the involved mathematical probabilities beforehand.

Still, too few players stop contributing to the lottery, even when they hear or find what the real probability figures are.

In a *6 from 49* lottery system, the probability of winning the 1st category with a single played variant (six numbers) is 1/13983816.

If playing weekly during a lifetime (let us assume eighty years of playing, respectively, 4320 draws), the probability for that player to finally win improves to 1/3236.

Still assuming that the person plays ten or even 100 variants once, he or she has a probability of 1/323 or 1/32, which is still low for a lifetime. And we did not even take into account the amount invested.

What exactly makes lottery players persevere in playing by ignoring these figures?

Beyond the addiction problems, there is also a psychological motivation of reference to community, having observation as a unique criterion.

A regular player may ask himself or herself the question: "If people all around me win the lottery, why can't I have my day once, too?"

Probability theory cannot completely answer to that question, but in exchange it can answer the question why that player has not won until the present moment: because the probability of winning is very close to zero.

Another example of false intuition is still related to the lottery.

Most players avoid playing the variant 1, 2, 3, 4, 5, 6.

Their argument is intuitive: It is impossible for the first six numbers to be drawn.

Indeed, it is almost impossible, in the sense that the probability of drawing that variant is 1/13983816.

Still, this probability remains the same for any played variant (assuming the technical procedure of drawing is absolutely random).

There are no preferential combinations, so that particular variant has not at all an inferior status from point of view of possibility of occurrence.

Moreover, if someone won by playing that variant, the amount won would be much higher than in the case of other played variant, because the winning fund will be divided (eventually) among fewer players.

Thus, the optimal decision would be to play that particular variant instead of others. Of course, this decision remains optimal as long as most players are not acquainted with this information.

False intuition successfully manifests in several gaming situations in gambling.

The so-called feeling of the player, which at a certain moments will influence a gaming decision, is very often a simple illusory psychical reaction that is not analytically grounded.

Probability represents one of the domains in which intuition may play bad tricks, even for persons with some mathematical education.

Therefore, intuition must not be used as a calculus tool or for probability estimation.

A correct probability calculus must be based on minimal, but clear, mathematical knowledge and must follow the basic logical algorithm of the application process, starting with framing the problem, then establishing the probability field and the calculus itself.

This process is described in detail in chapter titled *Beginner's Calculus Guide.*

PROBABILITY THEORY BASICS

In this chapter we present the main set of notions and foundation results for the mathematical concept of probability and probability theory.

Because this guide is addressed principally to beginners, we have limited it to the notions leading to the rigorous definition of probability and the properties generating the formulas that are necessary to practical calculus as well, especially for discrete and finite cases.

Denotation convention

In this chapter as well as in the following chapters that contain solved applications, the entire range of denotations corresponding to a specific operation or definition are used without their being limited.

For example, for the operation of multiplication, we use the symbols "×", "·" or no sign (in case of algebraic products that contain letters); for the operation of division we use the symbols

":", "−" or "/"; and for convergence we use the denotations

$$\text{"}\lim_{n \to \infty} a_n = a\text{", "}a_n \xrightarrow{n \to \infty} a\text{", "}a_n \xrightarrow{n} a\text{", etc.}$$

Fundamental notions

Sets

The set concept is a primary one in the sense that it cannot be defined through other mathematical notions.

In mathematics, the word *set* represents any well-defined collection of objects of any type (in the sense that we can decide whether a certain object belongs to respective collection or not) called elements of the set.

Specifying a set means enumerating its constituent objects or indicating a specific property of these objects (a common property that other objects do not have).

The sets are denoted in uppercase letters and the description of their elements is enclosed in braces.

Examples: $A = \{x, y, z\}$; $B = \{2,3,5,8\}$; $C = \{x \in R, 1 \le x \le 3\}$.

In certain theoretical constructions, we refer to objects that belong to a certain class of elements as a base set or reference set, usually denoted by Ω.

Example: The set of real valued numbers R can be considered a base set for its subsets: the set of integer numbers Z, of natural numbers N, of positive real numbers R_+.

A set can contain a finite or infinite number of elements. A set with no elements is called an *empty set* and is denoted by ϕ.

Example: The set of natural satellites of the Moon is an empty set.

If A is a set, B is called a *subset* of A if any element of B is also an element of A.

If Ω is a base set, we denote by $\mathcal{P}(\Omega)$ the *set of all parts* of Ω.

So $\mathcal{P}(\Omega)$ contains a number of parts A, B, C, ... that are individually well defined as sets.

Therefore, the elements of $\mathcal{P}(\Omega)$ are subsets of Ω.

Two sets are said to be *equal* if they contain the same elements.

Consider the following operations on $\mathcal{P}(\Omega)$:

1. The *union* of two sets A and B, denoted by $A \cup B$, is the set of elements that belong to A or B.

Example: If $A = \{2, 3, 5, 7\}$ and $B = \{3, 5, 9, 11\}$, then $A \cup B = \{2, 3, 5, 7, 9, 11\}$.

2. The *intersection* of two sets A and B, denoted by $A \cap B$, is the set of elements that belong to both A and B.

In the previous example, $A \cap B = \{3, 5\}$.

If $A \cap B = \phi$, we call the sets A and B *mutually exclusive* or *disjoint*.

3. The *complement* of A, denoted by A^C, is the set of elements from Ω that do not belong to A. Obviously, $A \cup A^C = \Omega$.

4. The *difference* of sets A and B, denoted by $A - B$ or A / B, is the set of elements from A that do not belong to B.

In the above example, $A - B = \{2, 7\}$ and $B - A = \{9, 11\}$.

Say that set A is *included* in set B (it is a subset of B) and denote it by $A \subset B$ (or $B \supset A$), if any element from A is an element of B.

The following properties, which are intuitively obvious, are characteristic for $\mathcal{P}(\Omega)$:

1. $\Omega \in \mathcal{P}(\Omega)$, $\phi \in \mathcal{P}(\Omega)$.
2. $A \in \mathcal{P}(\Omega) \Rightarrow A^C \in \mathcal{P}(\Omega)$.
3. $A, B \in \mathcal{P}(\Omega) \Rightarrow A \cup B \in \mathcal{P}(\Omega)$.
4. $A, B \in \mathcal{P}(\Omega) \Rightarrow A \cap B \in \mathcal{P}(\Omega)$.

Functions

Definition: Let A and B be two sets and let f be a correspondence law that associates a unique element from B to each element of A.

Call *function* defined on A with values in B the triple (A, f, B).

A is called the *domain* of the function, B is called the *range* of the function and f is called the *plot* (law) of the function (f is a part of the Cartesian product $A \times B$, which is the set of all pairs (a, b), with $a \in A$ and $b \in B$).

Instead of the denotation (A, f, B) the preferred denotation is:
$f : A \rightarrow B$ or $x \rightarrow f(x)$ or $y = f(x)$.

x is called a *variable* or argument.

The unique element $y \in B$ that corresponds to $x \in E$ is called the *image* of x through f or the value of function f in x.

If $B \subseteq R$ (the range is a subset of the real number system), the function is said to be a *real-valued function*.

Examples:

1) $A = \{a,b,c\}$, $B = \{d,e\}$, $f(a) = d$, $f(b) = e$, $f(c) = d$
f is a correspondence law that represents a function from A to B because it associates a unique element from B to each element of A.

2) $f : R \rightarrow R$, $f(x) = 2x$ is a function defined on the real number system, associating each real number with its double.

3) $g : [-1, 1] \rightarrow [0, 1]$, $g(x) = x^2$ associates each number from the interval $[-1, 1]$ with its quadrate.

4) Function $h : R \rightarrow Z$, $h(x) = [x]$ associates each real number with its integer part.

Definition: A function $f : A \rightarrow B$ is said to be *injective* if $\forall x, y \in A$ and $x \neq y$, then $f(x) \neq f(y)$.

In other words, a function is injective if any two different elements from the domain have different images.

Examples:
The function from Example 2) is injective.
Function g from Example 3) is not injective because the different elements -1 and 1 have the same image through g, that is, 1.
Also, the functions from Examples 1) and 4) are not injective.

Definition: A function $f : A \rightarrow B$ is said to be *surjective* if $\forall y \in B, \exists x \in A$ such that $f(x) = y$.

Or, written another way, $B = f(E)$, which means that the image of set A through function f is set B.

So, a function is surjective if any element from its range is the image of an element from its domain.

In other words, the function f is surjective if for any $b \in B$, the equation $f(x) = b$ has at least one solution $x \in E$. If, in addition, this solution is unique, the function f is also injective.

Examples:
All functions from the previous Examples 1) through 4) are surjective.

If we change the range of function h from Example 4) by substituting R for Z, function $h : R \rightarrow R$, $h(x) = [x]$ is no longer surjective because, if we take a noninteger number from R, it cannot be the integer part (the image through h) of a real number.

Definition: A function is said to be *bijective* if it is injective and surjective. A bijective function is also called a bijection.

Examples:
The only bijective function from Examples 1) through 4) is the one from Example 2).

Definition: A set A is said to be *countable* if there exists a bijection from A to the set of natural numbers N.

Examples:

1) The set of natural even numbers P is countable because there exists the bijective function $f : P \to N$, $f(x) = \dfrac{x}{2}$.

2) The set of natural uneven numbers I is countable because there exists the bijective function $g : I \to N$, $g(x) = \dfrac{x-1}{2}$.

3) The set of perfect squares (0, 1, 4, 9, 16, ...) is countable because the function $h(x) = \sqrt{x}$ is a bijection from this set to N.

Theorem
Any infinite set contains a countable subset.

Obviously, any countable set is infinite.
But, not every infinite set is countable.

Boole algebras

Definition: Call *Boole algebra* a nonempty set $\mathcal{A}$, with operations $\cup$, $\cap$, ^{C} defined and meeting the following axioms:

1. $A \cup B = B \cup A$; $A \cap B = B \cap A$ (commutativity);
2. $A \cup (B \cup C) = (A \cup B) \cup C$; $A \cap (B \cap C) = (A \cap B) \cap C$ (associativity);
3. $(A \cap B) \cup A = A$; $A \cap (A \cup B) = A$ (absorption);
4. $A \cap (B \cup C) = (A \cap B) \cup (A \cap C)$; $A \cup (B \cap C) = (A \cup B) \cap (A \cup C)$ (distributivity);
5. $(A \cap A^{C}) \cup B = B$; $(A \cup A^{C}) \cap B = B$ (complementarity), for any $A, B, C \in \mathcal{A}$.

Examples:

1) The set of parts $\mathcal{P}(\Omega)$ of a nonempty set Ω, with union, intersection and complement (related to Ω) operations, gets a Boole algebra structure.

2) The pair of residue classes of integers n, modulo 2, $L_2 = \{0, 1\}$ (the residues of division by 2), structured with operations:

$(x \cup y = x + y - xy, \; x \cap y = xy)$, is a Boole algebra.

From this definition it follows that:

Consequence 1 (duality transformation): If into a true statement containing $\cup$, $\cap$, ^{C} operations and $\subset$, $\supset$ relations, the following replacements are made throughout: $\cup$ with $\cap$, $\cap$ with $\cup$, $\subset$ with $\supset$ and $\supset$ with $\subset$, and ^{C} is left unchanged, we also obtain a true statement, called the *dual statement*.

Consequence 2 (idempotency laws): For any $A \in \mathcal{A}$, we have $A \cup A = A$, $A \cap A = A$.

Consequence 3 (monotony laws): For any $A, B, C \in \mathcal{A}$, if $A \subset B$, then $A \cup C \subset B \cup C$, $A \cap C \subset B \cap C$.

Consequence 4: For any sets $(A_i)_{1 \leq i \leq n} \subset \mathcal{A}$, the elements $A_1 \cup \ldots \cup A_n$ and $A_1 \cap \ldots \cap A_n$ are uniquely determined and do not depend on the order of operated elements.

Consequence 5: In a Boole algebra two elements exist, called the *null element*, denoted by Λ, and the *total element*, denoted by V, so that for any $A \in \mathcal{A}$, the following equalities are true:
$A \cap A^C = \Lambda$ and $A \cup A^C = V$.

Consequence 6: For any $A \in \mathcal{A}$, we have:
$A \cap V = A$, $A \cup V = V$, $A \cap \Lambda = \Lambda$, $A \cup \Lambda = A$ or $\Lambda \subset A$, $A \subset V$.

Consequence 7: If $A \cap B = \Lambda$ and $A \cup B = V$, then $B = A^C$.

Consequence 8 (de Morgan relations): For any $A, B \in \mathcal{A}$, we have: $(A \cup B)^C = A^C \cap B^C$ and $(A \cap B)^C = A^C \cup B^C$.

Sequences of real numbers. Limit

Definition: Let A be a set. Call a *sequence* of elements from A a function $f : N \to A$, where N is the set of natural numbers.

In other words, a sequence is an infinite enumeration of elements from A.

Although the set of values of function f is a subset of A, we do not say the sequence associated to it is a subset of A because it might contain repeated elements.

If we write $f(n) = a_n$ ($n \in N$, $a_n \in A$), we can also use the denotation $(a_n)_{n \geq 0}$ for this sequence.

a_k is called the *k-rank* term of the sequence.

Examples:

1) $f(n) = n$ is the sequence of natural numbers:
$(n)_{n \geq 0} = (1, 2, 3, \dots)$, $a_n = n$

2) $f(n) = 2n + 1$ is the sequence of natural uneven numbers:
$(2n + 1)_{n \geq 0} = (1, 3, 5, \dots)$, $a_n = 2n + 1$

3) A radioactive substance halves its mass every 12 hours.
By measuring the mass every 12 hours, the measurement results will be $M, \dfrac{M}{2}, \dfrac{M}{4}, \dfrac{M}{8}, \dots, \dfrac{M}{2^n}, \dots$

This is the sequence $\left(\dfrac{M}{2^n}\right)_{n \geq 0}$. Observe that its terms progressively approach zero, but without taking this value.

4) Let $A_1 = \{1\}$, $A_2 = \{1, 2\}$, $A_3 = \{1, 2, 3\}$, ...,
$A_n = \{1, 2, 3, \dots, n\}$, ...

The sequence $f(n) = A_n$ or $(A_n)_{n \geq 0}$ is a sequence of sets having the property $A_n \subset A_{n+1}$ for every n.

Definition: The sequences $(a_n)_{n \geq 0}$ and $(b_n)_{n \geq 0}$ are said to be *equal* if $a_n = b_n, \forall n \geq 0$.

If the set A from the sequence's definition is N, we call that sequence a sequence of natural numbers.
  If $A = Z$, we call it a sequence of integer numbers.
  If $A = Q$, we call it a sequence of rational numbers.
  If $A = R$, we call it a sequence of real numbers.

Definition: The sequence of real numbers $(a_n)_{n \geq 0}$ is said to be *monotonic increasing* (respectively *strictly increasing*), if $a_n \leq a_{n+1}$ (respectively $a_n < a_{n+1}$) for every n.
  The sequence of real numbers $(a_n)_{n \geq 0}$ is said to be *monotonic decreasing* (respectively *strictly decreasing*), if $a_n \geq a_{n+1}$ (respectively $a_n > a_{n+1}$) for every n.

If $a_n = a_{n+1}$ for every n, call the sequence *constant*.

Definition: The sequence of real numbers $(a_n)_{n \geq 0}$ is said to be *upper bounded* if a real valued number M exists such that $a_n \leq M$ for every n.
  The sequence of real numbers $(a_n)_{n \geq 0}$ is said to be *lower bounded* if a real number m exists such that $m \leq a_n$ for every n.
  The sequence of real numbers $(a_n)_{n \geq 0}$ is said to be *bounded* if has both upper and lower bounds (all the sequence's terms are placed within a bounded interval).
  The sequence of real numbers $(a_n)_{n \geq 0}$ is said to be *unbounded* if it is not bounded.

Examples:

1) The sequence $(3n + 2)_{n \geq 0}$ is strictly increasing.

2) The sequence $\left(\dfrac{1}{n} \right)_{n \geq 0}$ is strictly decreasing. This sequence is bounded, because $0 \leq \dfrac{1}{n} \leq 1$ for every n.

We now formally add two elements to the real line R, called *minus infinity* (denoted by $-\infty$) and *plus infinity* (denoted $+\infty$ or ∞).

Consider the set $\overline{R} = R \cup \{-\infty, \infty\}$, called the *closed real line*.

We can induce on $\overline{R}$ a relation order that extends the order from R, by putting $-\infty < \infty$, $-\infty < x$ and $x < \infty$ for any $x \in R$.

Definition: Let $a \in R$. Call a *vicinity* of a any set $V \subset R$ that contains an open interval that has a as its center (that is, $r > 0$ exists such that $(a - r, a + r) \subset V$).

Call a vicinity of ∞ (respectively $-\infty$) any set $V \subset \overline{R}$ that contains a $(b, \infty]$ (respectively $[-\infty, b)$) form interval, where b is a real valued number.

Examples:

1) The intervals $(-3, 3)$, $[-2, 3)$ or $\left(-\infty, \dfrac{5}{2} \right)$ are vicinities of point O (the origin), because they contain the open interval $(-1, 1)$, which is centered in the origin. The set Z is not a vicinity of the origin.

2) An open interval (a, b) is a vicinity of any of its own points.

A closed interval $[a, b]$ is a vicinity of any of its own points, except its heads, a and b.

129

Definition: Let $D \subset R$ be a subset of real numbers.

A point $\alpha \in \overline{R}$ is called an *accumulation point* for D if in any vicinity of α at least one point from $D - \{\alpha\}$ exists.

Examples:

1) For $D = (a, b)$, any point from $[a, b]$ is an accumulation point.

2) For $D = N$, ∞ is the only accumulation point.

3) For $D = (-\infty, 1) \cup (1, \infty)$, any real valued number α, including 1, is an accumulation point. $-\infty$ and ∞ are also accumulations points for D.

4) The set $D = (-1, 2] \cup \{3\}$ has the points of the closed interval $[-1, 2]$ as accumulation points, while 3 is not an accumulation point for D.

Definition: Let $(a_n)_{n \geq 0}$ be a sequence of real numbers and $a \in \overline{R}$. The sequence $(a_n)_{n \geq 0}$ is said to have the limit a if any vicinity of a contains all the sequence's terms from a certain rank upward.

This is written as $\lim\limits_{n \to \infty} a_n = a$ or $a_n \to a$ while $n \to \infty$

(*n* approaches infinity).

Intuitively, the sequence (a_n) has the limit a if its terms are accumulating around a or come closer to a as their rank approaches infinity:

$$a_0 \quad a_1 \quad a_2 \ldots\ldots\ldots a_n \ldots\ldots\ldots\ldots\ldots a \ldots\ldots$$

Definition: A sequence of real numbers that has a finite limit is called *convergent*. If $a \in R$ and $\lim_{n \to \infty} a_n = a$, the sequence $(a_n)_{n \ge 0}$ is said to be convergent toward a.

The sequences having no limit and those having the limit $+\infty$ or $-\infty$ are said to be *divergent*.

Examples:

1) The sequence $a_n = \dfrac{1}{n}$ ($n \ge 1$) is convergent toward 0

($\lim_{n \to \infty} \dfrac{1}{n} = 0$), because in any vicinity V of 0 we choose (let $(-\varepsilon, \varepsilon) \subset V$), all terms of a_n sequence are in V from the rank $\left[\dfrac{1}{\varepsilon}\right] + 1$ upward (in which we denoted by $[x]$ the integer part of x).

Indeed, $n \ge \left[\dfrac{1}{\varepsilon}\right] + 1 \Rightarrow n > \dfrac{1}{\varepsilon} \Rightarrow \dfrac{1}{n} < \varepsilon$.

2) The sequence $a_n = n^2$ has the limit $+\infty$, so it is divergent. If V is a random vicinity of ∞, $\varepsilon > 0$ exists such that $(\varepsilon, \infty) \subset V$. Put $N = [\sqrt{\varepsilon}] + 1$. It follows that $\forall n \ge N$, we have $n > \sqrt{\varepsilon}$, so $n^2 > \varepsilon$; therefore, all terms of a_n sequence are in V from rank N upward. This means that $\lim_{n \to \infty} n^2 = \infty$.

Theorem (the uniqueness of the limit)
If a sequence of real numbers has a limit, then this limit is unique.

Definition: Let $s = (a_n)_{n \ge 0}$ be a sequence of real numbers. If $k_0 < k_1 < k_2 < ... < k_n < ...$ is a strictly increasing sequence of natural numbers, then the sequence $(a_{k_n})_{n \ge 0}$ is called a *subsequence* of s.

Example: By putting $k_n = 2n$ we obtain the subsequence $(a_{2n})_{n \ge 0}$ of terms of even rank of s ($a_0, a_2, a_4, ...$).

Convergence and boundedness properties:

Theorem
Any convergent sequence of real numbers is bounded.

Theorem
Any subsequence of a convergent sequence is still convergent toward the same limit.

Theorem
a) Any sequence of real numbers that is monotonic increasing and upper bounded is convergent.
b) Any sequence of real numbers that is monotonic decreasing and lower bounded is convergent.

Cesaro's Lemma
Any bounded sequence of real numbers has at least one convergent subsequence.

Operations with convergent sequences:

Lemma
If $(u_n)_{n \geq 0}$ and $(v_n)_{n \geq 0}$ are sequences convergent toward zero, then, for any $\alpha, \beta \in R$ we have $\alpha u_n + \beta v_n \to 0$.

Theorem
Assume the sequences $(a_n)_{n \geq 0}$ and $(b_n)_{n \geq 0}$ are convergent.
Then, the sequences $(a_n + b_n)_{n \geq 0}$ and $(a_n \cdot b_n)_{n \geq 0}$ are convergent and, in addition,

$$\lim_{n \to \infty}(a_n + b_n) = \lim_{n \to \infty} a_n + \lim_{n \to \infty} b_n,$$
$$\lim_{n \to \infty}(a_n \cdot b_n) = \left(\lim_{n \to \infty} a_n\right) \cdot \left(\lim_{n \to \infty} b_n\right).$$

Theorem

If $(a_n)_{n \geq 0}$ and $(b_n)_{n \geq 0}$ are convergent sequences and $\lim_{n \to \infty} b_n \neq 0$, then the sequence $\left(\dfrac{a_n}{b_n} \right)_{n \geq k}$ is convergent and, in addition,

$$\lim_{n \to 0} \frac{a_n}{b_n} = \frac{\lim_{n \to \infty} a_n}{\lim_{n \to \infty} b_n}.$$

Theorem

If $(a_n)_{n \geq 0}$ and $(b_n)_{n \geq 0}$ are sequences of real numbers, both having a limit, and $a_n \leq b_n$ for every $n \geq N$ (where N is a fixed natural number), then $\lim_{n \to \infty} a_n \leq \lim_{n \to \infty} b_n$ (inequalities are preserved when they are passed to the limit).

Theorem (of tongs)

Suppose $(a_n)_{n \geq 0}$, $(b_n)_{n \geq 0}$ and $(c_n)_{n \geq 0}$ are three sequences of real numbers such that $a_n \leq b_n \leq c_n$ for every $n \geq N$ (where N is a fixed natural number).

If $a_n \to l$ and $c_n \to l$, then the sequence $(b_n)_{n \geq 0}$ has a limit and this is equal to l.

Series of real numbers

Let $(a_n)_{n \geq 1}$ be a sequence of real numbers and let $s_n = a_1 + a_2 + a_3 + \ldots + a_n$, for every $n \geq 1$.

The numbers s_n form in turn a sequence $(s_n)_{n \geq 1}$, which is called the *sequence of partial sums* of sequence $(a_n)_{n \geq 1}$.

The couple of sequences $\left((a_n)_{n \geq 1}, (s_n)_{n \geq 1} \right)$ is called a *series of real numbers* and is also denoted by $\sum_{n \geq 1} a_n$.

Definition: If the sequence of partial sums $(s_n)_{n \geq 1}$ is divergent, then the series is said to be *divergent*.

If the sequence of partial sums has a limit, then this limit is denoted by $\sum_{n=1}^{\infty} a_n$ and is called the *sum of the series*.

If this limit is finite, the series is said to be *convergent*.

If the sequence of partial sums has no limit, then the series is said to be *oscillating*.

Examples:

1) The geometric series $a + aq + aq^2 + ... + aq^{n-1} + ...$ is convergent if $-1 < q < 1$ and its sum is $\dfrac{a}{1-q}$.

2) The alternating harmonic series

$$1 - \frac{1}{2} + \frac{1}{3} - \frac{1}{4} + ... + (-1)^{n+1} \cdot \frac{1}{n} + ...$$ is convergent and its sum is *ln 2*.

3) The series $1 + 1 + 1 + ...$ is divergent and has the sum $+\infty$.

4) The series $1 - 1 + 1 - 1 + ...$ is oscillating.

Theorem

If $\sum_{n \geq 1} a_n$ is a convergent series of real numbers, then the sequence $(a_n)_{n \geq 1}$ converges toward zero.

Measure theory basics

In 1898, Emil Borel defined the concept of a measurable set of real numbers and developed a measure theory in order to solve some problems of complex analysis.

Georg Cantor, the founder of set theory, continued the line Borel mapped out.

We owe him the well-known result from elementary topology: any open set of real numbers is a countable union of open and mutually exclusive intervals.

Because of this result, we can measure the open sets as well as the closed sets from the topology of R.

We know that the length of an interval on the real axis can be represented by a real number, namely, the difference of abscissas of its terminal points.

The idea naturally follows of extending the notion of the length of an interval to sets that are much more complicated than the intervals.

Measure theory builds a function with certain properties, called measure, which generalizes the length notion.

The classical theory of measure and integral was defined by Henri Lebesgue at the beginning of 20th century in the Euclidean space R.

Measure concept was later generalized and abstracted.

Measure theory is a major tool that is indispensable for performing profound studies of probability theory and mathematical statistics.

Sequences of sets

Definition: Let X be a nonempty set. Call sequence of sets a function $f : N \to \mathcal{P}(X)$.

If we put $f(n) = A_n$, the sequence can also be denoted by $(A_n)_{n \in N}$.

Definition: The sets of a sequence $(A_n)_{n \in N}$ are said to be *mutually exclusive* if $A_i \cap A_j \neq \phi$ for any natural numbers $i \neq j$.

Definition: A sequence of sets $(A_n)_{n \in N}$ is said to be *increasing* or *ascending* (respectively *decreasing* or *descending*) if $A_n \subseteq A_{n+1}$ for every $n \in N$.

An increasing or decreasing sequence is called a *monotonic* sequence.

Definition: For a sequence of sets $(A_n)_{n \in N}$, the set of elements x from X that belong to an infinity of sets from the sequence is called the *upper limit* of the sequence and is written as $\overline{\lim} A_n$ or $\lim \sup A_n$.

The set of elements from X that belongs to all sets from the sequence, except for a finite number of them, is called the *lower limit* of the sequence and is written as $\underline{\lim} A_n$ or $\lim \inf A_n$.

From the definition of the two limits, it follows that $\underline{\lim} A_n \subseteq \overline{\lim} A_n$.

Definition: The sequence $(A_n)_{n \in N}$ is said to be convergent, if $\underline{\lim} A_n = \overline{\lim} A_n$. In this case, we write $A = \lim_{n \to \infty} A_n = \underline{\lim} A_n = \overline{\lim} A_n$.

The set A is called the limit of sequence $(A_n)_{n \in N}$.

Theorem

The upper and lower limits of a sequence of sets $(A_n)_{n \in N}$ are given by the relations:

$$\overline{\lim} A_n = \bigcap_{n=1}^{\infty} \bigcup_{k=n}^{\infty} A_k \,, \quad \underline{\lim} A_n = \bigcup_{n=1}^{\infty} \bigcap_{k=n}^{\infty} A_k \,.$$

Theorem

Any increasing sequence of sets $(A_n)_{n \in N}$ is convergent and its limit is the union of all sets.

Any decreasing sequence of sets $(A_n)_{n \in N}$ is convergent and its limit is the intersection of all sets.

Tribes. Borel sets. Measurable space

Definition: Let X be a set and $\mathcal{A}$ be a nonempty family of subsets of X. If $\mathcal{A}$ is a Boole algebra, then $\mathcal{A}$ is also called an *algebra of sets.*

From axioms 1 through 5 of the definition of Boole algebra and their consequences it follows that in an algebra of sets, the finite unions and intersections are well defined (they are uniquely determined as algebra's elements and do not depend on the order of operated sets):

If $(A_i)_{1 \le i \le n} \subset \mathcal{A}$, then $\displaystyle\bigcup_{i=1}^{n} A_i = A_1 \cup ... \cup A_n$ and

$\displaystyle\bigcap_{i=1}^{n} A_i = A_1 \cap ... \cap A_n$ are uniquely determined elements of $\mathcal{A}$.

Proposition

Let Ω be a family of subsets of the set X. Then exists the smallest algebra $\mathcal{A}$ containing Ω (that means, if $\mathcal{B}$ is an algebra containing Ω, then $\mathcal{A} \subseteq \mathcal{B}$).

Definition: Call the algebra generated by Ω and denoted by $\mathcal{A}(\Omega)$ the smallest algebra containing Ω.

Definition: Call σ-algebra, or *tribe*, an algebra $\mathcal{T}$ of sets that has the countable additivity property: the union of any countable family of sets from $\mathcal{T}$ is an element of $\mathcal{T}$:

$$A_n \in \mathcal{T}, n = 1, 2, 3, \ldots \Rightarrow \bigcup_{n=1}^{\infty} A_n \in \mathcal{T}.$$

Consequence

For every family $\{A_i\}_{i \in N}$ of elements from $\mathcal{T}$, we have $\bigcap_{i=1}^{\infty} A_i \in \mathcal{T}$

(because $\bigcap_{i=1}^{\infty} A_i = \left(\bigcup_{i=1}^{\infty} A_i \right)^C$).

The definition of tribe extends the operations of finite union and intersection on a Boole algebra to the countable union and intersection.

Examples:

1) $\mathcal{P}(X)$ (the set of parts of X) is a σ-algebra on X.

2) Let X = N. The set $\mathcal{T} = \{\phi, N, \{1, 3, 5, 7, \ldots\}, \{2, 4, 6, 8, \ldots\}\}$ is a σ-algebra on N.

Proposition
Let Ω be a family of subsets of set X. Then exists the smallest σ-algebra containing Ω.

Definition: Call σ-algebra generated by Ω and denote by $\sigma(\Omega)$ or $\mathcal{T}(\Omega)$, the smallest σ-algebra containing Ω.

We now extend the union and intersection operations for an arbitrary family $\mathcal{F}$ (not necessarily countable) of elements from a Boole algebra, as follows:

138

Definition: Call the union of elements $A \in \mathcal{F}$, the element $B \in \mathcal{F}$, if B meets these conditions:
1) $A \subset B$ for every $A \in \mathcal{F}$,
2) If $A \subset D$ for any $A \in \mathcal{F}$, then $B \subset D$.

Call the intersection of elements $A \in \mathcal{F}$, the element $C \in F$, if C meets these conditions:
3) $C \subset A$ for every $A \in \mathcal{F}$,
4) If $D \subset A$ for any $A \in \mathcal{F}$, then $D \subset C$.
Write $B = \bigcup_{A \in \mathcal{F}} A$ and $C = \bigcap_{A \in \mathcal{F}} A$.

If $F = \{A_i\}_{i \in I}$, where I is an arbitrary set of indexes, then the following denotations are used: $B = \bigcup_{i \in I} A_i$ and $C = \bigcap_{i \in I} A_i$.

Definition: Let X be a nonempty set. Call a *topology* on X a family of subsets of X, denoted by τ, that meets the following three axioms:
1) $\phi, X \in \tau$
2) $D_1, D_2 \in \tau$ Implies $D_1 \cap D_2 \in \tau$
3) If $D_i \in \tau$ for any $i \in I$, then $\bigcup_{i \in I} D_i \in \tau$.

The couple (X, τ) is called a *topological space*.
The elements of τ are called *open sets*.
A set A is said to be *closed* (in the τ topology) if its complement with respect to X is an open set.

Examples:

1) Let X be a nonempty set and $\tau_d = \mathcal{P}(X)$. The topological space (X, τ_d) is called discrete space and the topology τ_d is called a discrete topology.

2) On the set of real numbers R, structured with the usual order, we can naturally build a topology in which the open sets are the open intervals.

139

Properties of open and closed sets:

a) A set D from the topological space X is open if, and only if, for any $x \in D$ there exists an open set D_x such that $x \in D_x \subseteq D$.

b) A finite intersection of open sets from a topological space is still an open set (this does not remain true in the case of an arbitrary intersection).

c) ϕ and X are open and closed sets at the same time.

d) A finite union of closed sets is still a closed set.

e) An arbitrary intersection of closed sets is still a closed set.

Definition: Call the tribe of Borel sets on X and denoted by $\mathcal{B}(X)$ the tribe generated by the topology τ of X:
$$\mathcal{B}(X) = \mathcal{T}(\tau) \text{ or } \mathcal{B}(X) = \sigma(\tau).$$
The elements of the tribe $\mathcal{B}(X)$ are called *Borel sets*.

Proposition
The tribe $\mathcal{B}(X)$ of Borel sets is identical to the σ-algebra generated by the family $\mathcal{F}$ of the closed sets from X.

Definition: Call measurable space a couple $(X, \mathcal{T})$, where $\mathcal{T}$ is a tribe on X. The elements of $\mathcal{T}$ are called *measurable sets*.

Measure

In this section we define the notion of measure that is used to define probability on a σ-field of events.

It is called a positive measure to differentiate it from real, complex or spectral measures.

Definition: Let $(X, \mathcal{T})$ be a measurable space. Call positive measure on X a function $\mu : \mathcal{T} \to [0,\infty]$ that has the following properties:

1) μ is a countable additive (or σ-additive), that means, for any sequence $(E_i)_{i \in \mathbb{N}}$ of elements from $\mathcal{T}$, which are mutually exclusive, we have:

$$\mu\left(\bigcup_{i=1}^{\infty} E_i\right) = \sum_{i=1}^{\infty} \mu(E_i).$$

2) There exists at least one set $E_0 \in \mathcal{T}$ such that $\mu(E_0) < \infty$.

The triplet $(X, \mathcal{T}, \mu)$ is called a *measure space*.

We can use the shorter term *measure* instead of *positive measure*.

Examples:

1) Let $X = \mathbb{N}$, $\mathcal{T} = \mathcal{P}(X)$ and $\mu : \mathcal{T} \to [0,\infty]$, defined by:

$\mu(E) =$ the number of elements of E, *if E is a finite set* or ∞, *if* E *is an infinite set.*

The measure μ is called the *counting measure*.

2) Let $X \neq \phi$, $x_0 \in X$ and the function $\delta_{x_0} : \mathcal{P}(X) \to \mathbb{R}$, defined by $\delta_{x_0}(E) = 1$, *if* $x_0 \in E$ or 0, *if* $x_0 \notin E$.

δ_{x_0} is a measure called the *Dirac measure*.

Properties of positive measures:

Let $(X, \mathcal{T}, \mu)$ be a measure space.

a) $\mu(\phi) = 0$

b) The finite additivity of μ:

For any finite family $(E_i)_{i=1}^{n}$ of mutually exclusive sets from $\mathcal{T}$,

we have $\mu\left(\bigcup_{i=1}^{n} E_i\right) = \sum_{i=1}^{n} \mu(E_i)$.

c) The monotony of μ:

If E and F are two sets from $\mathcal{T}$ such that $E \subset F$, then $\mu(E) \leq \mu(F)$. If, in addition, $\mu(E) < \infty$, then $\mu(F-E) = \mu(F) - \mu(E)$.

d) The countable subadditivity of μ:

If $(E_i)_{i \in N}$ is a sequence of sets from $\mathcal{T}$, we have

$$\mu\left(\bigcup_{i=1}^{\infty} E_i\right) \leq \sum_{i=1}^{\infty} \mu(E_i).$$

e) Convergence properties:

Theorem

Let $(X, \mathcal{T}, \mu)$ be a measure space. If $(E_n)_{n \in N}$ is a sequence of sets from $\mathcal{T}$ such that $E_1 \subseteq E_2 \subseteq \ldots \subseteq E_n \subseteq E_{n+1} \subseteq \ldots$, then $\lim_{n \to \infty} \mu(E_n) = \mu\left(\lim_{n \to \infty} E_n\right)$.

Theorem

Let $(X, \mathcal{T}, \mu)$ be a measure space. If $(E_n)_{n \in N}$ is a sequence of sets from $\mathcal{T}$ such that $E_1 \supseteq E_2 \supseteq \ldots \supseteq E_n \supseteq E_{n+1} \supseteq \ldots$ and $\mu(E_1) < \infty$, then $\lim_{n \to \infty} \mu(E_n) = \mu\left(\lim_{n \to \infty} E_n\right)$.

Theorem

Let $(X, \mathcal{T}, \mu)$ be a measure space.

1) For any sequence $(E_n)_{n \in \mathbb{N}}$ of sets from $\mathcal{T}$, we have
$\mu(\underline{\lim}E_n) \le \underline{\lim}\mu(E_n)$.

2) If $\mu\left(\bigcup_{n=1}^{\infty} E_n\right) < \infty$, then $\mu(\overline{\lim}E_n) \ge \overline{\lim}\mu(E_n)$.

3) If the sequence $(E_n)_{n \in \mathbb{N}}$ is convergent and $\mu\left(\bigcup_{n=1}^{\infty} E_n\right) < \infty$, then
$\lim_{n \to \infty}\mu(E_n) = \mu\left(\lim_{n \to \infty} E_n\right)$.

Theorem of uniqueness of measures

Let $(X, \mathcal{T})$ be a measurable space, $\mathcal{M} \subset \mathcal{P}(X)$ such that $\mathcal{M}$ is closed to finite intersections, $E \in \mathcal{M}$ and $\mathcal{T} = \sigma(\mathcal{M})$, μ and ν two measures on $(X, \mathcal{T})$ such that $\mu = \nu$ on $\mathcal{M}$. Then, $\mu = \nu$.

Field of events. Probability

Probability theory deals with the laws of evolution of random phenomena.

Here are some examples of random phenomena:

1. The simplest example is the experiment involving rolling a die; the result of this experiment is the number that appears on the upper side of the die.

Even though we repeat the experiment several times, we cannot predict which value each roll will take because it depends on many random elements like the initial impulse of the die, the die's position at the start, characteristics of the surface on which the die is rolled, and so on.

2. A person walks from home to his or her workplace each day. The time it takes to walk that distance is not constant, but varies because of random elements (traffic, meteorological conditions, and the like).

3. We cannot predict the percentage of misfires when firing a weapon a certain number of times at a target.

4. We cannot know in advance what numbers will be drawn in a lottery.

In these experiments, the essential conditions of each experiment are unchanged.

All variations are caused by secondary elements that influence the result of the experiment.

Among the many elements that occur in the phenomena studied here, we focus only on those that are decisive and ignore the influence of secondary elements.

This method is typical in the study of physical and mechanical phenomena as well as in technical applications.

In the study of these phenomena, there is a difference of principle between the methods that allow the essential elements that determine the main character of the phenomenon to be taken into account and those methods that do not ignore the secondary elements that lead to errors and perturbations.

The randomness and complexity of causes require special methods of study of random phenomena, and these methods are elaborated by probability theory.

The application of mathematics in the study of random phenomena is based on the fact that, by repeating an experiment many times in identical conditions, the relative frequency of a certain result (the ratio between number of experiments having one particular result and total number of experiments) is about the same, and oscillates around a constant number.

If this happens, we can associate a number with each event; that is, the probability of that event.

This link between structure (the structure of a field of events) and number is the equivalent of the mathematics of the transfer of quality into quantity.

The problem of converting a field of events into a number is equivalent to defining a numeric function on this structure, which has to be a measure of the possibility of an event occurring.

Because the occurrence of an event is probable, this function is named *probability*.

Probability theory can only be applied to phenomena that have a certain stability of the relative frequencies around probability (homogeneous mass phenomena).

This is the basis of the relationship between probability theory and the real world and daily practice.

So, the scientific definition of probability must first reflect the real evolution of a phenomenon.

Probability is not the expression of the subjective level of man's trust in the occurrence of the event, but the objective characterization of the relationship between conditions and events, or between cause and effect.

The probability of an event makes sense as long as the set of conditions is left unchanged; any change in these conditions changes the probability and, consequently, the statistical laws governing the phenomenon.

The discovery of these statistical laws resulted from a long process of abstraction.

Any statistical law is characterized, on the one hand, by the relative inconstancy or the variability of various objects' activity, and therefore we cannot predict the evolution of an individual object. On the other hand, for a large set of phenomena a stable constancy takes place, and this can be expressed by the statistical law.

Practical statistics works first with finite fields of events, while physical and technical experiments take place on infinite fields of events.

In probability theory, the studied events result from random *experiments* (trials) and each performance of an experiment is called a *test*.

The result of a test is called an *outcome*. An experiment can have more than one outcome, but any test has a single outcome.

An *event* is a set of outcomes.

As a result of a test with outcome e, an event A *occurs* if $e \in A$ and *does not occur* if $e \notin A$.

Example:

In the roll of a die, the set of all possible outcomes is $\Omega = \{1, 2, 3, 4, 5, 6\}$. Some events are: $A = \{1, 3, 5\}$ – uneven number, $B = \{1, 2, 3, 4\}$ – less than the number 5, $C = \{2, 4, 6\}$ – even number.

If rolling a 3, events A and B occur.

Denote by Ω the set of all possible outcomes of an experiment and by $\mathcal{P}(\Omega)$ the set of all parts of Ω.

Ω is called the set of outcomes or the *sample space*.

The random events are elements of $\mathcal{P}(\Omega)$.

On the set Σ of the events associated with an experiment, we can introduce three operations that correspond to the logical operations *or, and, non*. Let $A, B \in \Sigma$.

a) *A or B* is the event that occurs if, and only if, one of the events A or B occurs. This event is denoted by $A \cup B$ and is called the union of events A and B.

b) *A and B* is the event that occurs if, and only if, both events A and B occur. This event is denoted by $A \cap B$ and is called the intersection of events A and B.

c) *non A* is the event that occurs if, and only if, event A does not occur. This event is called the complement (opposite) of A and is denoted by A^c.

If we attach to each event the set of tests through which it occurs, then the operations between events revert to the respective operations between sets of corresponding tests, so the designations a), b) and c) are justified.

The results of the operations with events are also events attached to their respective experiments.

If $A \cap B = \phi$, meaning A and B cannot occur simultaneously, we say that A and B are *incompatible* (mutually exclusive) events.

If $A \cup B = \Omega$, we say that A and B are *collectively exhaustive*.

In the set Σ of events associated with a certain experiment, two events with special significance exist, namely, event $\Omega = A \cup A^C$ and event $\phi = A \cap A^C$.

The first consists of the occurrence of event A or the occurrence of event A^C, which obviously always happens; that means this event does not depend on event A. It is natural to call Ω the *sure event*.

Event ϕ consists of the occurrence of event A and the occurrence of event A^C, which can never happen. This event is called the *impossible event*.

Let $A, B \in \Sigma$. We say that event A implies event B and write $A \subset B$, if, when A occurs, B necessarily occurs.

If we have $A \subset B$ and $B \subset A$, we say that events A and B are equivalent and write $A = B$ (this reverts to the equality of the sets of tests that correspond to respective events).

The implication between events is a partial order relation on the set of events and corresponds to the inclusion relation from Boole algebras.

Definition: An event $A \in \Sigma$ is said to be *compound* if two events $B, C \in \Sigma$, $B \neq A$, $C \neq A$ exist, such that $A = B \cup C$. Otherwise, the event A is said to be *elementary*.

If the set Ω contains a finite number of elementary events, $\Omega = \{\omega_1, \omega_2, ..., \omega_k\}$, then an event is part of Ω and therefore it contains a finite number $(r < k)$ of elementary events.

In this case, the set of events Σ is $\mathcal{P}(\Omega)$ itself.

Usually, we have $\Sigma \subset \mathcal{P}(\Omega)$ and Ω has the same properties as $\mathcal{P}(\Omega)$.

We have the following:

Axiom
The set of events associated to an experiment is a Boole algebra.

Definition: The Boole algebra of the events associated to an experiment is called the *field of events* of that experiment.

So, the field of events is a set of parts of Ω, structured as an algebra of events Σ, and is denoted by $\{\Omega, \Sigma\}$.

Examples:

1) An urn contains twenty balls numbered 1 to 20. A ball is drawn and its number is memorized.
   a) Write the sure event.
   b) For some events, let A – the number is even; B – the number is a multiple of 5; and C – the number is a power of 2.

   Write the events $A \cup B$, $A \cap B$, A^C. Show the implications between events. Which events are incompatible?

Answer:
a) $\Omega = \{1, 2, 3, 4, 5, 6, 7, 8, 9, 10, 11, 12, 13, 14, 15, 16, 17, 18, 19, 20\}$.
b) $A \cup B$ – the number is even or multiple of 5.
   $A \cap B$ – the number is multiple of 10.
   A^C – the number is uneven.
   $C \subset A$ and $C \cap B = \phi$.

2) In the die rolling experiment, write the field of events associated to this experiment and calculate the number of events of this field.

Answer:
   We have $\{\Omega, \Sigma\} = \{\phi, \{i\}, \{i, j\}, \{i, j, k\}, \{i, j, k, l\}, \{i, j, k, l, m\}, \Omega\}$,

   Where i, j, k, l, m take independent values from 1 to 6, and all indexes are different within a subgroup.

   Two subgroups with the same number of indexes differ from each other by at least one index.

We denoted by $\{i\}$ – rolling number i, by $\{i, j\}$ - rolling number i or number j (so, $\{i, j\} = \{i\} \cup \{j\}$), etc.

Events $\{i\}$ are in number of C_6^1, $\{i, j\}$ in number of C_6^2, $\{i, j, k\}$ in number of C_6^3, etc. Thus, in this field we have

$$1 + C_6^1 + C_6^2 + C_6^3 + C_6^4 + C_6^5 + C_6^6 = 2^6 \text{ events. (See the}$$

combinations' properties)

3) An urn contains four white balls w_1, w_2, w_3, w_4 and two black balls b_1, b_2. Two balls are simultaneously drawn.
   a) Specify the tests of this experiment.
   b) Consider the events:
       A_1 – drawing two black balls,
       A_2 – drawing two white balls,
       A_3 – drawing at least one black ball,
       A_4 – drawing a single white ball,
       A_5 – drawing a single black ball,
       A_6 – drawing two green balls.

Specify which of them are random, elementary or compound, the pairs of compatible and incompatible events, and the implications between them.

Answer:
a) Denote symbolically by (w_1, w_2) the draw of balls w_1 and w_2, etc. The tests of this experiment are:

$$(w_1, w_2), (w_1, w_3), (w_1, w_4), (w_2, w_3), (w_2, w_4), (w_3, w_4),$$
$$(w_1, b_1), (w_1, b_2), (w_2, b_1), (w_2, b_2), (w_3, b_1), (w_3, b_2),$$
$$(w_4, b_1), (w_4, b_2), (b_1, b_2).$$

The number of tests is $C_6^2 = 15$.

b) Events A_1, A_2, A_3, A_4, A_5 are random.

Event A_6 is impossible; it is neither elementary nor compound.

Event A_1 is elementary, it occurs through a single test, (b_1, b_2).

We have $A_1 = \{(b_1, b_2)\}$.

Events A_2, A_3, A_4, A_5 are compound.

We have $A_4 = A_5$, because the occurrence of A_4 implies the occurrence of A_5 and vice versa. This can be also observed in the equality of sets of tests:

$A_4 = \{(w_1, b_1), (w_1, b_2), (w_2, b_1), (w_2, b_2), (w_3, b_1),$
$(w_3, b_2), (w_4, b_1), (w_4, b_2)\}$.

$A_5 = \{(w_1, b_1), (w_2, b_1), (w_3, b_1), (w_4, b_1), (w_1, b_2), (w_2, b_2),$
$(w_3, b_2), (w_4, b_2)\}$.

We have:

$A_2 = \{(w_1, w_2), (w_1, w_3), (w_1, w_4), (w_2, w_3), (w_2, w_4),$
$(w_3, w_4)\}$.

$A_3 = \{(w_1, b_1), (w_1, b_2), (w_2, b_1), (w_2, b_2), (w_3, b_1),$
$(w_3, b_2), (w_4, b_1), (w_4, b_2), (b_1, b_2)\}$.

The pairs (A_3, A_5), (A_1, A_3), (A_4, A_5) are compatible; these events can occur simultaneously.

The pairs (A_2, A_3), (A_1, A_2), (A_1, A_4), (A_1, A_3), (A_2, A_4), (A_2, A_5) are incompatible.

Events A_1 and A_4, A_1 and A_5, A_2 and A_4, A_2 and A_3, A_2 and A_5 are opposite one to another.

We have the implications: $A_4 \subset A_5$, $A_5 \subset A_4$, $A_4 \subset A_3$, $A_5 \subset A_3$.

We now present some additional properties of elementary events:

(E1) Let $A \in \Sigma$ be an elementary event and let $B \in \Sigma$ be an event. If $B \subseteq A$, then $B = \phi$ or $B = A$.

(E2) A necessary and sufficient condition for an event $A \in \Sigma$, $A \neq \phi$ to be elementary is an event $B \in \Sigma$, $B \neq \phi$ such that $B \subset A$ not to exist.

(E3) A necessary and sufficient condition for an event $A \neq \phi$ to be elementary is that for any event B to have $A \cap B = \phi$ or $A \cap B = A$.

(E4) Any two different elementary events are incompatible.

(E5) In a finite algebra of events, for any compound event $B \in \Sigma$, an elementary event A exists, such that $A \subset B$.

(E6) Any event from a finite algebra of events can be written in a unique form as a union of elementary events.

(E7) In a finite algebra of events, the sure event is the union of all elementary events.

Probability on a finite field of events

Let us consider an urn U that contains n balls, from which m balls are white and $n - m$ balls are black (the only difference among the balls is color). A ball is drawn at random.
We have n elementary events.
Let A be the event *the drawn ball is white.* This event can occur after m tests, $m \leq n$.

Definition: Call the probability of event A the ratio between the number of situations favorable for A to occur and the number of equally possible situations. Therefore, $P = \dfrac{m}{n}$.

This is the classical definition of probability.
It can only be used in experiments having equally possible elementary events.

Now, let us consider an urn with n balls, from which a_1 balls have color c_1, a_2 balls have color c_2, ... , a_s balls have color c_s.
So $n = a_1 + a_2 + ... + a_s$.
Again, the only difference among the balls is color.

151

A ball is drawn from the urn. In this case, the drawing of a ball is an elementary event.

The probability for a ball with color c_k to be drawn is given by the classical definition of probability: $P = \dfrac{a_k}{n}$.

So, a favorable event is to draw a ball with color c_k.

An event of the field is the appearance of a ball having color c_{k_1}, c_{k_2}, ... or c_{k_s}, denoted by A, and

$$P(A) = \frac{a_{k_1} + a_{k_2} + ... + a_{k_s}}{n}.$$

Let us observe that:

1. The probability of each event is a function of that event, with positive values.

2. The probability of the sure event (Ω) is 1.

$$P(\Omega) = \frac{a_1 + a_2 + ... + a_s}{n} = 1.$$

3. If $A = A_1 \cup A_2$, with $A_1 \cap A_2 = \phi$, then

$$P(A) = P(A_1) + P(A_2).$$

4. The elementary events are equally probable (each has the probability $\dfrac{1}{n}$).

The experiment of drawing balls from an urn can be interpreted by considering two fields of events: the field considered above, where an elementary event is to draw a ball and another field (Γ, $\mathcal{P}(\Gamma)$), where an elementary event is to draw a ball with color c_k ($k = 1, 2, ..., s$).

The function $P(A)$, that is, the probability for one event from Γ to occur, has the properties 1, 2, and 3, but generally does not have the property 4, because the elementary events from Γ do not have equal probabilities; this only happens when $a_1 = a_2 = ... = a_s$.

Let us consider an urn with seven balls numbered from 1 to 7, from which a ball is drawn at random.

Denoting by A_i, $i = 1, \ldots, 7$, the event of drawing ball number i (taking into consideration that the balls only differ from one another by their number labels), it follows that events A_i are equally possible, so, naturally, a probability function defined on the set $(A_i)_{1 \le i \le 7}$ meets the condition $P(A_i) = p$, with $p = \dfrac{1}{7}$.

The event $A_i \cup A_j$, $i \ne j$, is twice as probable as each of the events A_i; therefore, the extension of function P to event $A_i \cup A_j$ must be defined as $P(A_i \cup A_j) = \dfrac{2}{7}$ $(i \ne j)$, so

$$P(A_i \cup A_j) = P(A_i) + P(A_j).$$

Generally, $P(A_{i_1} \cup \ldots \cup A_{i_k}) = \dfrac{k}{7}$ $(1 \le i_1 \le \ldots \le i_k \le 7)$ and $P(\Omega) = 1$, where Ω is the sure event.

Therefore, in the case of a finite field of events $\{\Omega, \Sigma\}$, a probability on this field is defined as follows:

Definition: Call probability on Σ a function $P : \Sigma \to R$ that satisfies the following axioms:
(1) $P(A) \ge 0$, for any $A \in \Sigma$;
(2) $P(\Omega) = 1$;
(3) $P(A_1 \cup A_2) = P(A_1) + P(A_2)$, for any $A_1, A_2 \in \Sigma$ that $A_1 \cap A_2 = \phi$.

Property (3) can be generalized through recurrence to any finite number of mutually exclusive events.

Therefore, if $A_i \cap A_j = \phi$, $i \ne j$ $(i, j = 1, \ldots, n)$, then:

$$P(A_1 \cup A_2 \cup \ldots \cup A_n) = \sum_{i=1}^{n} P(A_i).$$

Example:

A bag contains 200 lotto tickets, among them one winning ticket of \$100, five winning tickets of \$50 each, ten winning tickets of \$20 each and twenty winning tickets of \$10 each. Someone buys a ticket. What is the probability of that person winning at least \$20?

Answer:

Consider the following events:

A - buyer wins at least \$20;

A_1 - buyer wins \$20;

A_2 - buyer wins \$50;

A_3 - buyer wins \$100.

We have $A = A_1 \cup A_2 \cup A_3$, $A_i \cap A_j = \phi$ ($i \neq j$, $i, j = 1, 2, 3$).

Therefore:

$$P(A) = P(A_1) + P(A_2) + P(A_3) = \frac{10}{200} + \frac{5}{200} + \frac{1}{200} = 0.08 = 8\%.$$

Definition: A finite field of events $\{\Omega, \Sigma\}$, structured with a probability P, is called a finite probability field (finite probability space) and is denoted by $\{\Omega, \Sigma, P\}$.

From the additivity rule of probability we deduce that, to find the probabilities of all events $A \in \Sigma$, it is sufficient to know the probabilities of the elementary events ω_i, $1 \leq i \leq r$ that the finite set $\Omega = \{\omega_1, ..., \omega_r\}$ consists of because, if $P(\{\omega_i\}) = p_i$, $1 \leq i \leq r$ and $A = \{\omega_{i_1}\} \cup ... \cup \{\omega_{i_k}\}$, then $P(A) = P(\{\omega_{i_1}\} \cup ... \cup \{\omega_{i_k}\}) = P(\{\omega_{i_1}\}) + ... + P(\{\omega_{i_k}\}) = p_{i_1} + ... + p_{i_k}$.

Hence, a finite probability field is completely characterized by the non-negative numbers $p_1, p_2, ..., p_r$, with $\sum_{i=1}^{r} p_i = 1$.

If $p_1 = ... = p_k$, then $P(A) = \dfrac{k}{r}$, where k is the number of elementary events that event A consists of (the elementary events that are favorable for event A to occur). In this way, we are driven to the classical definition of probability.

Probability properties

We have the following properties of the probability function:

(P1) For any $A \in \Sigma$, we have $P(A^C) = 1 - P(A)$.

(P2) $P(\phi) = 0$.

(P3) Any $A \in \Sigma$, $0 \le P(A) \le 1$.

(P4) Any $A_1, A_2 \in \Sigma$ with $A_1 \subset A_2$, we have $P(A_1) \le P(A_2)$.

(P5) Any $A_1, A_2 \in \Sigma$, we have
$P(A_2 - A_1) = P(A_2) - P(A_1 \cap A_2)$.

(P6) If $A_1, A_2 \in \Sigma$, $A_1 \subset A_2$, then $P(A_2 - A_1) = P(A_2) - P(A_1)$

(P7) Any $A_1, A_2 \in \Sigma$, we have
$P(A_1 \cup A_2) = P(A_1) + P(A_2) - P(A_1 \cap A_2)$.

(P8) Any $A_1, A_2 \in \Sigma$, we have $P(A_1 \cup A_2) \le P(A_1) + P(A_2)$.

(P9) If $(A_i)_{1 \le i \le n} \subset \Sigma$, then

$$P(A_1 \cup A_2 \cup ... \cup A_n) = \sum_{i=1}^{n} P(A_i) - \sum_{j<i} P(A_i \cap A_j) +$$

$$+ \sum_{i<j<k} P(A_i \cap A_j \cap A_k) + ... + (-1)^{n-1} P(A_1 \cap A_2 \cap ... \cap A_n)$$

This property is also called the *inclusion-exclusion principle*.

(P10) Let $(A_i)_{1 \le i \le n} \subset \Sigma$ be events, with
$P(A_1 \cap A_2 \cap ... \cap A_{n-1}) \ne 0$. Then:
$P(A_1 \cap A_2 \cap ... \cap A_n) = P(A_1)P(A_2 / A_1)P(A_3 / A_1 \cap A_2)...$
$P(A_n / A_1 \cap A_2 \cap ... \cap A_{n-1})$.

The previous properties represent formulas currently used in probability calculus on a finite field of events.

Property (P9) is the main calculus formula for applications in finite cases.

Applications:

1) Two shooters are simultaneously shooting one shot each at a target.

The probabilities of target hitting are 0.8 for the first shooter and 0.6 for the second.

Calculate the probability for the target to be hit by at least one shooter.

Answer:

Let A_i – *shooter number* i *(i = 1, 2) hits the target* be the events. The events A_1 and A_2 are independent, so

$P(A_1 \cap A_2) = P(A_1) \cdot P(A_2)$ (See the section titled *Independent events. Conditional probability*).

We have, according to (P7):

$P(A_1 \cup A_2) = P(A_1) + P(A_2) - P(A_1 \cap A_2) =$
$= 0.8 + 0.6 - 0.8 \times 0.6 = 0.92$.

2) A batch of 100 products is amenable to quality control.

The batch will be rejected if at least one defective product among five randomly controlled products is found.

Assuming the batch contains 4 percent defective products, calculate the probability for the batch to be rejected.

Answer:

Denoting by *A* the event *the batch must be rejected,* we calculate $P(A^C)$.

We denote by A_k the event *controlled product number* k *is accepted (not defective),* $1 \le k \le 5$. Events A_k are not independent.

We have:

$P(A^C) = P(A_1 \cap ... \cap A_5) = P(A_1)P(A_2 / A_1)...$

156

$$P(A_5 / A_1 \cap A_2 \cap A_3 \cap A_4) = \frac{96}{100} \cdot \frac{95}{99} \cdot \frac{94}{98} \cdot \frac{93}{97} \cdot \frac{92}{96} \text{ and}$$

$P(A) = 1 - P(A^C)$. (We have used (P10) and (P1))

3) An urn contains twenty balls numbered from 1 to 20.

If we choose four different numbers between 1 and 20 and randomly extract four balls from urn, we can calculate the probability for at least two extracted balls to have numbers from the chosen four.

Answer (to follow this solution, read the chapter titled *Combinatorics*):

We call a *variant* any group of four different extracted balls.

The whole number of possible variants is $C_{20}^4 = 4845$.

Without restricting the generality, we can denote the chosen numbers by 1, 2, 3, 4 (the probability is the same for any other four numbers).

A variant containing numbers 1 and 2 has a (1, 2, x, y) form, with x, y receiving different values ($x \neq y$) from $20 - 2 = 18$ numbers.

The number of combinations taken two at a time from the chosen four numbers is $C_4^2 = 6$, namely, (1, 2), (1, 3), (1, 4), (2, 3), (2,4), (3, 4). Let:

A_1 – the extracted variant contains the balls numbered 1 and 2,

A_2 – the extracted variant contains the balls numbered 1 and 3,

A_3 – the extracted variant contains the balls numbered 1 and 4,

A_4 – the extracted variant contains the balls numbered 2 and 3,

A_5 – the extracted variant contains the balls numbered 2 and 4,

A_6 – the extracted variant contains the balls numbered 3 and 4.

We can write, as sets:

$A_1 = \{(1, 2, x, y); x \neq y; x, y \in \{1, 2, ..., 20\} - \{1, 2\}\}$,

$A_2 = \{(1, 3, x, y); x \neq y; x, y \in \{1, 2, ..., 20\} - \{1, 3\}\}$,

$A_3 = \{(1, 4, x, y); x \neq y; x, y \in \{1, 2, ..., 20\} - \{1, 4\}\}$,

$A_4 = \{(2, 3, x, y); x \neq y; x, y \in \{1, 2, ..., 20\} - \{2, 3\}\}$,

$A_5 = \{(2, 4, x, y); x \neq y; x, y \in \{1, 2, ..., 20\} - \{2, 4\}\}$,

$A_6 = \{(3, 4, x, y); x \neq y; x, y \in \{1, 2, ..., 20\} - \{3, 4\}\}$.

Each of sets A_i $(i = 1, ..., 6)$ has $C_{18}^2 = 153$ elements (within a variant, the order of numbers does not matter).

We must calculate $P(A_1 \cup A_2 \cup ... \cup A_6)$.

To do this, we apply property (P9) (the inclusion-exclusion principle).

Observe that the probabilities standing in the (P9) formula, applied to our case, represent fractions with the same denominator (4845).

Therefore, the problem reverts to one of counting the elements of the sets:

$(A_i \cap A_j)_{1 \le i < j \le 6}$, $(A_i \cap A_j \cap A_k)_{1 \le i < j < k \le 6}$,

$(A_i \cap A_j \cap A_k \cap A_l)_{1 \le i < j < k < l \le 6}$,

$(A_i \cap A_j \cap A_k \cap A_l \cap A_m)_{1 \le i < j < k < l < m \le 6}$ and

$(A_1 \cap A_2 \cap A_3 \cap A_4 \cap A_5 \cap A_6)$.

1) *The intersections of the* A_i *sets taken two at a time*

We have $C_6^2 = 15$ intersections of the A_i sets taken two at a time.

In succession, simply stated, we denote the set of variants $\{(1, 2, x, y); x \ne y; x, y \in \{1, 2, ..., 20\} - \{1, 2\}\}$ by (12xy), and so on.

Let us consider the following three intersections:

$(12xy) \cap (13xy)$, $(12xy) \cap (14xy)$, $(13xy) \cap (14xy)$ – number 1 contained in each variant.

A variant that is common to sets (12xy) and (13xy) has a (123x) form, with x receiving $20 - 3 = 17$ values.

So, the intersection $(12xy) \cap (13xy)$ has 17 elements.

The other two intersections also have 17 elements.

We similarly count elements for the intersections containing one of the numbers 2, 3 and 4 in each variant (nine intersections).

So, we have $3 + 9 = 12$ intersections with 17 elements each.

The rest of $15 - 12 = 3$ intersections, namely $(12xy) \cap (34xy)$, $(13xy) \cap (24xy)$ and $(23xy) \cap (14xy)$, each contains a single element, respectively the variant (1, 2, 3, 4) (because all four numbers must be in the common variant).

Therefore, the total number of elements of all intersections of the A_i sets taken two at a time is $3 \times 1 + 12 \times 17 = 207$.

2) *The intersections of the* A_i *sets taken three at a time*

We have $C_6^3 = 20$ intersections.

From these, we have four intersections, respectively:

$(12xy) \cap (13xy) \cap (14xy)$ – number 1 in each variant,

$(12xy) \cap (23xy) \cap (24xy)$ – number 2 in each variant,

$(13xy) \cap (23xy) \cap (34xy)$ – number 3 in each variant,

$(14xy) \cap (24xy) \cap (34xy)$ – number 4 in each variant,

each has a single element, namely (1, 2, 3, 4).

The other four intersections are:

$(12xy) \cap (23xy) \cap (13xy)$ – two of the numbers 1, 2 and 3 in each variant,

$(23xy) \cap (34xy) \cap (24xy)$ – two of the numbers 2, 3 and 4 in each variant,

$(13xy) \cap (34xy) \cap (14xy)$ – two of the numbers 1, 3 and 4 in each variant,

$(12xy) \cap (24xy) \cap (14xy)$ – two of the numbers 1, 2 and 4 in each variant.

Each has 17 elements because a variant that is common to sets $(12xy)$, $(23xy)$ and $(13xy)$ has a $(123x)$ form, with x receiving $20 - 3 = 17$ values.

We count the elements of the other three intersections in the same way.

For the rest of $20 - 8 = 12$ intersections each has a single element, namely (1, 2, 3, 4) because each of numbers 1, 2, 3, 4 must be in one common variant.

Therefore, the total number of elements of all intersections of the A_i sets taken three at a time is $4 \times 17 + 16 \times 1 = 84$.

3) *The intersections of the* A_i *sets taken four at a time*

We have $C_6^4 = 15$ intersections.

Any variant of an intersection of four A_i sets must contain each of the numbers 1, 2, 3, 4.

This means that any intersection has a single element, namely the (1, 2, 3, 4) variant.

Therefore, the total number of elements of all intersections of the A_i sets taken four at a time is $1 \times 15 = 15$.

4) *The intersections of the A_i sets taken five at a time*

We have $C_6^5 = 6$ intersections.

Similar to case 3), we have a single element for each intersection; therefore, the total number of elements is $1 \times 6 = 6$.

5) *The intersection of all six A_i sets*

The intersection has a single element for the same reason as in previous case.

All cases have been covered and now we can apply the (P9) formula:

$$P(A_1 \cup A_2 \cup ... \cup A_6) = \frac{6 \cdot 153 - 207 + 84 - 15 + 6 - 1}{4845} = 0.16202.$$

Probability σ - field

The definition of the probability σ-field and of probability on such field corresponds to the Kolmogorov axiomatics of probability theory given by Andrei Nicolaevich Kolmogorov in 1933.

This definition extends the probability notion on a finite field of events.

Definition: Call a σ-field of events a field of events $\{\Omega, \Sigma\}$ that has the countable additivity property: any countable union of events from Σ is still an event from Σ (if $(A_n)_{n \geq 0} \subset \Sigma$, then $\bigcup\limits_{n=1}^{\infty} A_n \in \Sigma$).

The definition corresponds to the definition of tribe in measure theory.

Definition: Let $\{\Omega, \Sigma\}$ be a σ-field of events. Call probability on $\{\Omega, \Sigma\}$ a numerical positive function P, defined on Σ, which meets the following conditions:

1) $P(\Omega) = 1$

2) $P\left(\bigcup_{i=1}^{\infty} A_i\right) = \sum_{i=1}^{\infty} P(A_i)$, for any countable family of mutually exclusive events $(A_i)_{i=1}^{\infty} \subset \Sigma$.

A σ-field of events $\{\Omega, \Sigma\}$ along with a probability P is called a probability σ-field and is written $\{\Omega, \Sigma, P\}$.

Observe that the two conditions from the definition of probability imply the axioms in the definition of measure.

Therefore, this probability is a measure with $\mu(\Omega) = 1$, so it acquires all the properties of a measure.

The terms currently used in the measure theory and those used in probability theory correspond as follows:

Measure Theory	Probability Theory
Measurable space	Field of events
Tribe	σ-field of events
Measurable set	Event
Whole space	Sure event
Empty set	Impossible event
Measure space	Probability field
Measure of a set	Probability of an event

The properties of probability on a finite field also stand for the probability σ-fields (the properties (P1) to (P10) from the section titled *Probability on a finite field of events*).

In addition, if $\{\Omega, \Sigma, P\}$ is a probability σ-field, we also have the following properties:

(P11)   For any sequence of events $(A_n)_{n \geq 0} \subset \Sigma$ with $A_{n+1} \subseteq A_n$ (descending), we have $\lim\limits_{n \to \infty} P(A_n) = P\left(\bigcap\limits_{n=0}^{\infty} A_n\right)$.

For any sequence of events $(A_n)_{n \geq 0} \subset \Sigma$ with $A_{n+1} \supseteq A_n$ (ascending), we have $\lim\limits_{n \to \infty} P(A_n) = P\left(\bigcup\limits_{n=0}^{\infty} A_n\right)$.

(P12)   For any sequence of events $(A_n)_{n \geq 0} \subset \Sigma$, we have
$$P(\underline{\lim} A_n) \leq \underline{\lim} P(A_n) \leq \overline{\lim} P(A_n) \leq P(\overline{\lim} A_n).$$

(P13)   If the sequence of events $(A_n)_{n \geq 0} \subset \Sigma$ is convergent ($\overline{\lim} A_n = \underline{\lim} A_n$), then $P\left(\lim\limits_{n \to \infty} A_n\right) = \lim\limits_{n \to \infty} P(A_n)$.

(P14)   Generally, $P\left(\bigcup\limits_{n=0}^{\infty} A_n\right) \leq \sum\limits_{n=0}^{\infty} P(A_n)$. The equality stands only if the events are mutually exclusive.

Independent events. Conditional probability

Let us consider the experiment of tossing two coins and let
A – *heads on first coin* and B – *heads on second coin* be two events.

The occurrence of event *A* and its probability do not depend on the occurrence of event *B*, and vice versa.

In this case, events *A* and *B* are said to be independent (each is independent of the other).

Consider an urn containing four white balls and three black balls. Two people extract one ball each from the urn.

Let A_1 – *first person is extracting a white ball* and A_2 – *second person is extracting a white ball* be two events.

The probability of event A_2, in the absence of information about A_1, is $\dfrac{4}{7}$.

162

If event A_1 has occurred, the probability of event A_2 is $\dfrac{1}{2}$, so event A_2 depends on event A_1.

Therefore, these two events are not independent.

Definition: Events A and B from the probability field $\{\Omega, \Sigma, P\}$ are said to be *P- independent* if $P(A \cap B) = P(A) \cdot P(B)$.

Proposition

If events A and B from Σ are P-independent, then the pairs of events (A, B^C), (B, A^C) and (A^C, B^C) are P-independent.

Definition: Events $\left(A_i\right)_{1 \le i \le n} \subset \Sigma$ are P-independent *m* by *m* if, for $h \le m$ and $1 \le i_1 < i_2 < ... < i_h \le n$, we have:

$$P\left(A_{i_1} \cap ... A_{i_h}\right) = P\left(A_{i_1}\right) \cdot ... \cdot P\left(A_{i_h}\right).$$

If $m = n$, the events are fully P-independent.

If *n* events are mutually independent, they are not necessarily fully independent, as seen in the following example.

Example:

Let us consider a homogenous tetrahedron having three sides colored white, black, red and the fourth side colored with all three colors.

Assume we perform an experiment in which we throw this object one time.

Let us denote by A_i the event *the tetrahedron lands on the side number* i *(i = 1, 2, 3, 4)*. Events A_i are elementary events of the field associated to the described experiment and $P(A_i) = \dfrac{1}{4}$ for every *i*.

If we denote by A the event *the tetrahedron lands on the white color,* by B the event *the tetrahedron lands on the black color* and by C the event *the tetrahedron lands on the red color,* then:

$A = A_1 \cup A_4$, $B = A_2 \cup A_4$ and $C = A_3 \cup A_4$.

We have $P(A) = P(B) = P(C) = 1/2$, because there are four possible and two favorable cases for landing on each color—the side with that respective color and the side with all colors.

On the other hand,

$$P(A \cap B) = P(B \cap C) = P(A \cap C) = P(A_4) = 1/4.$$

It follows that events A, B and C are mutually P-independent.

From $P(A \cap B \cap C) = P(A_4) = 1/4$ and $P(A)\,P(B)\,P(C) = 1/8$ it results that events A, B and C are not fully P-independent.

Definition: Events $(A_n)_{n \geq 0} \subset \Sigma$ are said to be P-independent if the events are P-independent in any finite number group.

In the urn example, we saw that the result of the experiment that generates A_1 influences the conditions of the experiment that generates A_2 (if the first person extracts a white ball from the urn, only three white balls and three black balls remain in the urn from which the second person can extract a ball).

Hence, the probability of event A_2 is influenced by the occurrence of A_1, in the sense that this probability is 1/2 if A_1 occurred and 2/3 if A_1 did not occurred.

It is natural to call the probability of event A_2 conditional on event A_1 and to denote it by $P(A_2|A_1)$ or $P_{A_1}(A_2)$.

Definition: Let $\{\Omega, \Sigma, P\}$ be a probability σ-field and $A, B \in \Sigma$ with $P(B) \neq 0$. Call the probability of event A conditional on event B, the ratio $P(A|B) = \dfrac{P(A \cap B)}{P(B)}$.

Proposition

If $\{\Omega, \Sigma, P\}$ is a probability σ-field and $B \in \Sigma$, then the triplet $\{\Omega, \Sigma, P_B\}$ is a probability σ-field.

Similarly, we can define the probability of event B as conditional on event A: $P(B|A) = \dfrac{P(A \cap B)}{P(A)}$, where $P(A) \neq 0$.

From the two analogue relations results

$$P(A)P(B|A) = P(B)P(A|B).$$

From the definition results that, in the case of two independent events, we have $P(A|B) = P(A)$.

Total probability formula. Bayes's theorem

Definition: Call a *complete system of events* a finite or countable family of events $(A_i)_{i \in I}$, with $A_i \cap A_j = \phi$ for any $i \neq j$, $i, j \in I$ and

$$\bigcup_{i \in I} A_i = \Omega.$$

Theorem (the total probability formula)

Let $(A_i)_{i \in I} \subset \Sigma$ be a complete system of events with $P(A_i) \neq 0, \forall i \in I$. For any $A \in \Sigma$, we have $P(A) = \sum_{i \in I} P(A_i) P(A|A_i)$.

Application:

Three urns contain the following numbers of balls: the first contains three white balls and two black balls; the second contains two white balls and one black ball; and the third contains four white balls and five black balls.

An urn is randomly chosen and a ball is drawn.

Find the probability for the drawn ball to be white.

Solution:

Denote by A_i – the urn number i is chosen (i = 1, 2, 3) and let A be the event *a white ball is drawn*.

We have $P(A_1) = P(A_2) = P(A_3) = 1/3$.

We also have $P(A|A_1) = 3/5$, $P(A|A_2) = 2/3$, $P(A|A_3) = 4/9$.

According to the total probability formula, we have:

$$P(A) = P(A_1) P(A|A_1) + P(A_2) P(A|A_2) + P(A_3) P(A|A_3) = 77/135.$$

Bayes's Formula (the theorem of hypotheses)

Let $(A_i)_{i \in I} \subset \Sigma$ be a complete system of events. The probabilities of these events (hypotheses) are given before performing an experiment. The experiment produces another event A.

Bayes's formula shows how the occurrence of event A changes the probabilities of these hypotheses. This formula is:

$$P(A_i|A) = \frac{P(A_i) \cdot P(A|A_i)}{\sum_{i \in I} P(A_i) \cdot P(A|A_i)} \text{ , for every } i \in I.$$

The law of large numbers

The previous theoretical statements defined the concept of probability on a field of events and offered tools for basic probability calculus.

The result we present now is qualitative; in fact, it illustrates the way in which probability models the hazard.

We enunciate the Law of Large Numbers, not in its general mathematical form, in order to avoid having to define more complex concepts, but in an exemplified particular form, in a way that everyone can understand.

The particular enunciation is the following classic result, known as Bernoulli's Theorem:

The relative frequency of the occurrence of a certain event in a sequence of independent experiments performed under identical conditions converges toward the probability of that event.

The theorem states that if A is an event, (E_n) a sequence of independent experiments, a_n the number of occurrences of event A after the first n experiments, then the sequence of non-negative numbers $\left(\dfrac{a_n}{n} \right)$ is convergent and its limit is $P(A)$:

The expression a_n is called *frequency* and the expression $\dfrac{a_n}{n}$ is called *relative frequency*.

We exemplify this expression by considering the classical experiment of tossing a coin:

Let A be the event *the coin falls heads up*. Obviously, $P(A) = 1/2$.

Let us say that event A has the following occurrences:

– after the first throw, 0 occurrences, relative frequency 0/1
– after the first two throws, 0 occurrences, relative frequency 0/2
– after the first three throws, 1 occurrence, relative frequency 1/3
– after the first four throws, 1 occurrence, relative frequency 1/4
– after the first five throws, 2 occurrences, relative frequency 2/5

....

– after the first n throws, a_n occurrences, relative frequency $\dfrac{a_n}{n}$.

The Law of Large Numbers says that the sequence 0, 0, 1/3, 1/4, ..., a_n / n, ... is convergent toward 1/2.

In other words, while n is growing, the relative frequency is approximating 1/2 with higher accuracy.

Or we can say we can find a big enough number of experiments for the relative frequency of occurrences of A to approach 1/2 with any decimal place we want.

The Law of Large Numbers confers on probability a property of limit.

Of course, the theorem does not provide information about the terms of the sequence of relative frequencies, but only about its limit.

In other words, we cannot make a precise prediction on the event at a certain chronological moment, but we can only know the behavior of the relative frequency of occurrences at infinity.

The general form of the theorem is presented in another section.

Discrete random variables

One of fundamental notions of probability theory is the notion of *random variable.*

The events of a probability field are not in principle values in the numerical sense of the term.

However, they can be described with the help of some real valued numbers that usually result from a measurement.

The definition of random variables allows the direct use of the values the experiment provides as a result, both in practice and in the theories intended to interpret that result as well.

If by random variable we mean a real valued function defined on the set of elementary events associated with a considered experiment, we can use examples that are specific to probability theory to illustrate how we can turn from an event to a random variable.

Let us consider an experiment having event *A* as result.

Instead of event *A*, we can consider the random variable *X* that takes the value 1 if *A* occurs and 0 if A^C occurs.

Thus, we have defined a Bernoullian random variable with two values (also called an indicator variable of event *A*) through the relation:

$$X(\omega) = \begin{cases} 1, \text{ if } \omega \in A \\ 0, \text{ if } \omega \in A^C \end{cases}$$

This scheme can be extended in the sense that we can consider an experiment having as its result a complete system of events $(A_i)_{1 \le i \le n}$.

If we assign the value $n - j$ in the case of event A_j occurring, $1 \le j \le n$, we define a Bernoullian random variable *X* with *n* values, through the relation: $X(\omega) = n - j$, if $\omega \in A_j, 1 \le j \le n$.

Let $\{\Omega, \Sigma, P\}$ be a σ-probability field and $(A_i)_{i \in I} \subset \Sigma$ a complete system of events (finite or countable).

The numerical system $p_i = P(A_i)$, $i \in I$ is called the *distribution* of the σ-probability field.

Definition: Call a *discrete random variable* a function X defined on the set of elementary events $\omega \in \Omega$ with real values if

1) X takes the values $x_i, i \in I$;

2) $\{\omega : X(\omega) = x_i\} \in \Sigma, i \in I$.

A discrete random variable in which I is finite is called *simple random variable*.

Schematically, the random variable X is denoted by:

$$X : \begin{pmatrix} x_i \\ p_i \end{pmatrix}, \ i \in I, \sum_{i \in I} p_i = 1.$$

The table corresponding to this denotation is called distribution or *repartition* of the random variable X.

Examples:

1) The random variable representing the number resulting from rolling the die has the following repartition:

$$X : \begin{pmatrix} 1 & 2 & 3 & 4 & 5 & 6 \\ 1/6 & 1/6 & 1/6 & 1/6 & 1/6 & 1/6 \end{pmatrix}.$$

2) A batch of pieces is the subject of quality control in this way:
Each piece is extracted consecutively from the batch and checked.

Five pieces are checked. If the piece from the extraction with rank $k = 1, 2, 3, 4$ is not good, the batch is rejected.

Write the repartition table of the random variable representing the number of checked pieces, if the probability of a randomly chosen piece being accepted is 0.85.

Answer:
The random variable X can take the values:

1 – if the first extracted piece is not good; we have $P(X = 1) = 0.15$;

2 – if the first extracted piece is good and the second is not; we have $P(X = 2) = 0.85 \cdot 0.15 = 0.1275$;

3 – if the first two extracted pieces are good and the third is not; we have $P(X = 3) = (0.85)^2 \cdot 0.15 = 0.108375$;

4 – if the first three extracted pieces are good and the fourth is not; $P(X = 4) = (0.85)^3 \cdot 0.15 = 0.0921187$;

5 – if the first four pieces are good; we have
$$P(X = 5) = (0.85)^4 = 0.5220062.$$

The repartition table is:

$$X : \begin{pmatrix} 1 & 2 & 3 & 4 & 5 \\ 0.15 & 0.1275 & 0.108375 & 0.0921187 & 0.5220062 \end{pmatrix}.$$

Conversely, if X is a discrete random variable that can take the values x_i ($i = 1, 2, ...$), we can assign to it a corresponding complete system of events $(A_i)_{i=1,2,...}$ such that $A_i = \{\omega : X(\omega) = x_i\}$.

Let X and Y be two random variables defined through:

$X(\omega) = x_n$, for $\omega \in A_n$, $n = 1, 2, ...$

$Y(\omega) = y_n$, for $\omega \in B_m$, $m = 1, 2, ...$,

$\{A_n\}$ and $\{B_m\}$ being two complete systems of events.

Definition: The random variables X and Y are said to be *independent* if for any m and n we have:
$P(A_n \cap B_m) = P(A_n) \cdot P(B_m)$, or, in other words, the complete systems of events $\{A_n\}$ and $\{B_m\}$ are independent.

Let X and Y be two arbitrary random variables defined through the above relations and let f be a real valued function of two real variables.

The random variable $Z = f(X, Y)$ takes the values $z_{nm} = f(x_n, y_m)$ and is defined by the complete system of events $C_{nm} = \{\omega : X(\omega) = x_n, Y(\omega) = y_m\}$.

If X and Y are independent, we have:
$P(X = x_n, Y = y_m) = P(X = x_n)P(Y = y_m)$.

170

Using the denotations $p_n = P(X(\omega) = x_n)$, $q_m = P(Y(\omega) = y_m)$,

results $P(Z = z) = \sum\limits_{f(x_n, y_m)} p_n q_m$.

For $f(x,y) = x + y$ we have $P(Z = z) = \sum\limits_{x_n + y_m = z} p_n q_m$.

If, in particular, X and Y can only take integer non-negative

values, we have $P(Z = k) = \sum\limits_{j=0}^{k} p_j q_{k-j}$.

The distribution $Z = X + Y$ is called the composition or *sum* of the random variables X and Y.

As an example, consider the following simple random variables:

$$X : \begin{pmatrix} x_1 & \cdots & x_n \\ p_1 & \cdots & p_n \end{pmatrix} \text{ and } Y : \begin{pmatrix} y_1 & \cdots & y_m \\ q_1 & \cdots & q_m \end{pmatrix}.$$

The random variable $X + Y$ has the following distribution table:

$$X + Y : \begin{pmatrix} x_1 + y_1 & x_1 + y_2 & \cdots & x_i + y_j & \cdots & x_n + y_m \\ p_{11} & p_{12} & \cdots & p_{ij} & \cdots & p_{nm} \end{pmatrix}, \text{ where}$$

$p_{ij} = P(X + Y = x_i + y_j) = P(\{\omega : X(\omega) = x_i\} \cap \{\omega : Y(\omega) = y_j\})$,

with $\sum\limits_{i=1}^{n} \sum\limits_{j=1}^{m} p_{ij} = 1$.

In the first row of the table, all possible combinations $x_i + y_j$ have been generated.

If X and Y are independent, we have $p_{ij} = p_i q_j$.

The *product* of random variables X and Y has the following distribution table:

$$XY : \begin{pmatrix} x_1 y_1 & x_1 y_2 & \cdots & x_i y_j & \cdots & x_n y_m \\ p_{11} & p_{12} & \cdots & p_{ij} & \cdots & p_{nm} \end{pmatrix}, \text{ where}$$

$p_{ij} = P(XY = x_i y_j) = P(\{\omega : X(\omega) = x_i\} \cap \{\omega : Y(\omega) = y_j\})$.

In the first row of the table, all possible combinations $x_i y_j$ have been generated.

The sum and product operations can be extended to any finite number of random variables.

The *power* of a random variable X has the following distribution table:

$$X^k : \begin{pmatrix} x_1^k & \cdots & x_n^k \\ p_1 & \cdots & p_n \end{pmatrix}, \text{ because } P(X^k = x_i^k) = P(X = x_i) = p_i,$$

for $i = 1, 2, ..., n$.

The *inverse* of a random variable with non-zero values has the

following distribution table: $X^{-1} : \begin{pmatrix} \dfrac{1}{x_1} & \cdots & \dfrac{1}{x_n} \\ p_1 & \cdots & p_n \end{pmatrix}$.

If random variable Y admits an inverse, then we can define the

quotient $\dfrac{X}{Y} = XY^{-1}$, which has the following distribution table:

$$\frac{X}{Y} : \begin{pmatrix} x_1 & \cdots & x_i & \cdots & x_n \\ y_1 & & y_j & & y_m \\ p_{11} & \cdots & p_{ij} & \cdots & p_{nm} \end{pmatrix}.$$

A constant can be interpreted as a random variable defined on a

set of elementary events, and its distribution table is $a : \begin{pmatrix} a \\ 1 \end{pmatrix}$.

We can now perform operations between random variables and constants.

Example:
Let $ax + by + c = 0$ be an equation in which the coefficients a, b and c are determined by rolling a die. Find the probability for the line thus obtained to run through the point $(-1, 1)$.

Answer:
The random variable X takes the numbers occurring on the die as values that can be attached to the experiment of determining a coefficient, so:

$$X : \begin{pmatrix} 1 & 2 & 3 & 4 & 5 & 6 \\ 1/6 & 1/6 & 1/6 & 1/6 & 1/6 & 1/6 \end{pmatrix}.$$

Let X_1, X_2, X_3 be the random variables corresponding to the determination of coefficients a, b, respectively c.

These random variables are independent.

The line runs through the point $(-1, 1)$ when $-X_1 + X_2 + X_3 = 0$.

Therefore, we must determine

$$P(-X_1 + X_2 + X_3 = 0) = P(X_1 = X_2 + X_3).$$

By taking into account that the random variables are independent, we can write:

$$P(-X_1 + X_2 + X_3 = 0) = \sum_{k=1}^{6} P(\{X_2 + X_3 = k\} \cap \{X_1 = k\}) =$$

$$\sum_{k=1}^{6} P(X_2 + X_3 = k)P(X_1 = k).$$

We have $P(X_1 = k) = \dfrac{1}{6}$, $k = 1, ..., 6$.

$\sum_{k=1}^{6} P(X_2 + X_3 = k) = P(X_2 + X_3 \leq 6)$ and this is the probability

for the sum of the points occurring on two dice to be less than 6.

The distribution table of the random variable that represents the sum of occurring points is:

$$\begin{pmatrix} 2 & 3 & 4 & 5 & 6 & 7 & 8 & 9 & 10 & 11 & 12 \\ \dfrac{1}{6^2} & \dfrac{2}{6^2} & \dfrac{3}{6^2} & \dfrac{4}{6^2} & \dfrac{5}{6^2} & \dfrac{6}{6^2} & \dfrac{5}{6^2} & \dfrac{4}{6^2} & \dfrac{3}{6^2} & \dfrac{2}{6^2} & \dfrac{1}{6^2} \end{pmatrix}.$$

We have then $\sum_{k=1}^{6} P(X_2 + X_3 = k) = \dfrac{15}{36}$.

It follows that $P(X_1 = X_2 + X_3) = \dfrac{5}{72}$.

173

Moments of a discrete random variable

Definition: Let X be a discrete random variable that takes the values x_i, with corresponding probabilities $p_i, i \in I$. If series $\sum_{i \in I} x_i p_i$ is convergent, the expression $M(X) = \sum_{i \in I} x_i p_i$ is called the *mean* of the discrete random variable X.

It is also called *mathematical expectation* or *expected value*.

If X is a simple random variable that takes the values $x_1, ..., x_n$, with corresponding probabilities $p_1, ..., p_n$, then its mean is

$$M(X) = \sum_{i=1}^{n} x_i p_i .$$

We now present a few additional properties of the mean of a random variable:

(M1) If X and Y are two discrete random variables and if $M(X)$ and $M(Y)$ exist, then the mean $M(X + Y)$ exists and we have $M(X + Y) = M(X) + M(Y)$.

(M2) Let $X_k, (k = 1, ..., n)$ be n discrete random variables. If $M(X_k)$ exists for every k, then $M\left(\sum_{k=1}^{n} X_k\right)$ exists and we have

$$M\left(\sum_{k=1}^{n} X_k\right) = \sum_{k=1}^{n} M(X_k).$$

(M3) Let X be discrete random variable and let c be a constant. If $M(X)$ exists, then $M(cX)$ exists and we have $M(cX) = cM(X)$.

(M4) Let $X_k, (k = 1, ..., n)$ be n discrete random variables and let $c_k, (k = 1, ..., n)$ be n constants. If $M(X_k)$ exists for every k, then $M\left(\sum_{k=1}^{n} c_k X_k\right)$ exists and we have $M\left(\sum_{k=1}^{n} c_k X_k\right) = \sum_{k=1}^{n} c_k M(X_k).$

(M5) The mean of random variable $X - M(X) = Y$ is null.
Y is called the *deviation* of the random variable X.

(M6) Schwartz's inequality: Let X and Y be two discrete random variables whose $M(X^2)$ and $M(Y^2)$ both exist. Then we have:
$$|M(XY)| \leq \sqrt{M(X^2)M(Y^2)}.$$

(M7) If X and Y are two independent discrete random variables and if $M(X)$ and $M(Y)$ both exist, then $M(XY)$ exists and we have $M(XY) = M(X)\,M(Y)$.

Examples:

1) An urn contains N balls numbered from 1 to N.
Let X be the highest number obtained as result of extracting n balls from the urn, provided the ball is put back in the urn after each extraction. Calculate $P(X = k)$ and $M(X)$.

Answer:
Let us denote by $\{X \leq k\}$ the event *the ball's number does not exceed the value k*, for each extraction.
Because the ball is put back in the urn after each extraction, it follows that the extractions are independent.
For each extraction, the probability for the ball's number to not exceed k is $\dfrac{k}{N}$; therefore, $P(X \leq k) = \left(\dfrac{k}{N}\right)^n$.

We then have:
$$P(X = k) = P(\{X \leq k\} - \{X \leq k-1\}) = P(X \leq k) - P(X \leq k-1) =$$
$$\left[k^n - (k-1)^n\right]\frac{1}{N^n}.$$

$$M(X) = \sum_{k=1}^{N} kp_k = \frac{1}{N^n}\sum_{k=1}^{N} k\left[k^n - (k-1)^n\right] =$$

$$\frac{1}{N^n}\sum_{k=1}^{N}\left[k^{n+1} - (k-1)^{n+1} - (k-1)^n\right] = N - \frac{1}{N^n}\sum_{k=1}^{N}(k-1)^n.$$

2) A device is formed by five components that may fail independently of each other. We number the components from 1 to 5 and let $p_k = 0.3 + 0.2(k-1)$ be the probability for component number k to fail ($k = 1, ..., 5$).

Calculate the mean of the number of failures.

Answer:

Let X_k be the random variable associated to component number k, which takes the values 1 and 0, depending on whether the component fails or not:

$$X_k : \begin{pmatrix} 1 & 0 \\ 0.3+0.2(k-1) & 0.7-0.2(k-1) \end{pmatrix}, \quad k = 1, ..., 5.$$

The random variable that gives us the number of failures is $X = X_1 + X_2 + ... + X_5$, so $M(X) = M(X_1) + M(X_2) + ... + M(X_5)$.

We have $M(X_k) = 0.3 + 0.2(k-1)$ and results

$$M(X) = \sum_{k=1}^{5} [0.3 + 0.2(k-1)] = 3.5.$$

Definition: Let X be a discrete random variable and let r be a natural number. If the mean of the random variable X^r exists, then this mean is called the *r-th moment* of random variable X and is denoted by $\alpha_r(x) = M(X^r) = \sum_k x_k^r p_k$.

The mean of random variable $|X|^r$ is called the *r-th absolute moment* of random variable X and is denoted by

$$\beta_r(X) = M(|X|^r) = \sum |x_k|^r p_k.$$

Definition: Given a discrete random variable X, the *r-th moment* of the random variable of the deviation of X is called the *centered r-th moment* of X and is denoted by $\mu_r(X) = \alpha_r(X - M(X))$.

The centered second moment of discrete random variable X is called the *dispersion* or *variance* and is denoted by $D^2(X)$ or σ^2, so $D^2(X) = \sigma^2 = \mu_2(X)$.

The number $D(X) = \sigma = \sqrt{\mu_2(X)}$ is called the *standard deviation* of X.

We now present a few additional properties of dispersion and standard deviation.

(D1) The following equality takes place:
$$D^2(X) = M(X^2) - [M(X)]^2.$$

(D2) If $Y = aX + b$, where a and b are constants, then
$$D(Y) = |a| \cdot D(X).$$

(D3) Let X_k, $(k = 1, ..., n)$ be n discrete random variables that are mutually independent and let c_k, $(k = 1, ..., n)$ be n constants.

We then have: $D^2\left(\sum_{k=1}^{n} c_k X_k\right) = \sum_{k=1}^{n} c_k^2 D^2(X_k).$

(D4) Chebyshev's inequality: Let X be a random variable.

We then have: $P(\{\omega : |X(\omega) - M(X)| \geq \varepsilon\}) < \dfrac{D^2(X)}{\varepsilon^2}$, for any $\varepsilon > 0$.

This inequality can be put in a form used frequently in applications: if we take $\varepsilon = aD(X)$, it can be written as

$$P(|X - M(X)| \geq aD(X)) < \frac{1}{a^2}.$$

Distribution function

Definition: Call the *distribution function* of the random variable X, the function $F(x) = P(\{\omega : X(\omega) < x\})$, defined for any $x \in R$.

From this definition we see that any random variable can be determined through its distribution function.

If X is a discrete random variable with $p_n = P(X = x_n)$, $n \in I$, then the expression from the definition can be written as $F(x) = \sum_{x_n < x} p_n$

and is called discrete distribution function.

In this case, F is a step function, which means it takes constant values on each interval determined by points x_i $(i \in I)$.

Example:

Let $X : \begin{pmatrix} 0 & 1 & 2 & 3 & 4 \\ 0.15 & 0.45 & 0.20 & 0.15 & 0.05 \end{pmatrix}$ be a discrete random

variable. We have:

$F(x) = 0$ if $x \le 0$;
$F(x) = 0.15$ if $0 < x \le 1$;
$F(x) = 0.15 + 0.45 = 0.60$ if $1 < x \le 2$;
$F(x) = 0.15 + 0.45 + 0.20 = 0.80$ if $2 < x \le 3$;
$F(x) = 0.15 + 0.45 + 0.20 + 0.15 = 0.95$ if $3 < x \le 4$.
$F(x) = 1$ if $x > 4$.

Theorem
The distribution function of a random variable has the following properties:
1) $F(x_1) \le F(x_2)$ if $x_1 < x_2$, $x_1, x_2 \in R$;
2) $\lim_{\substack{y \to x \\ y < x}} F(y) = F(x)$, $\forall x \in R$;
3) $\lim_{x \to -\infty} F(x) = 0$;
4) $\lim_{x \to +\infty} F(x) = 1$.

The reciprocal of this theorem is also true: any function $F(x)$ having properties 1) through 4) is the distribution function of a random variable defined on a conveniently chosen probability field.

Theorem
Let X be a random variable whose distribution function is $F(x)$ and let a and b be two real valued numbers with a < b.
The following equalities take place:
1) $P(a \le X < b) = F(b) - F(a)$;
2) $P(a < X < b) = F(b) - F(a) - P(X = a)$;
3) $P(a < X \le b) = F(b) - F(a) - P(X = a) + P(X = b)$;
4) $P(a \le X \le b) = F(b) - F(a) + P(X = b)$.

Classical discrete probability repartitions

Bernoulli scheme

Let us consider the following problem:
Three independent shots are fired at a target.
The probability of hitting the target is p for each of the three shots.
Find the probability for two of the three shots hitting the target.
Let A be the event *two of the three shots hit the target* and let A_i ($i = 1, 2, 3$) be the events *shot number* i *hit the target.*
A can be written as:

$$A = \left(A_1 \cap A_2 \cap A_3^C\right) \cup \left(A_1 \cap A_2^C \cap A_3\right) \cup \left(A_1^C \cap A_2 \cap A_3\right).$$

The three parentheses are incompatible and the events they consist of are independent.
This gives us the result:

$$P(A) = P(A_1)P(A_2)P(A_3^C) + P(A_1)P(A_2^C)P(A_3) + P(A_1^C)P(A_2)P(A_3)$$

therefore, $P(A) = 3p^2(1 - p)$.

We can solve the following more general problem using a similar approach:
Consider that n independent experiments are performed.
In each experiment, event A may occur with probability p and does not occur with probability $q = 1 - p$.
Find the probability for event A to occur exactly m times in the n experiments.

Let B_m be the event *A occurs exactly* m *times in the* n *experiments* and let A_i ($i = 1, 2, ..., n$) be the events *A did not occur in the* i-*th experiment.*

Each variant of occurrence of B_m consists of m occurrences of event A and of $n - m$ nonoccurrences of A (that is $n - m$ occurrences of A^C).
We then have:

$$B_m = \left(A_1 \cap A_2 \cap ... \cap A_m \cap A_{m+1}^C \cap ... \cap A_n^C\right) \cup$$
$$\left(A_1^C \cap A_2 \cap ... A_{m+1} \cap A_{m+2}^C \cap ... \cap A_n^C\right) \cup ... \cup$$
$$\left(A_1^C \cap ... \cap A_{n-m}^C \cap A_{n-m+1} \cap ... \cap A_n\right).$$

The number of ways we can choose m experiments in which A occurs from the n experiments is C_n^m.

All variants are incompatible and the experiments are independent, so:

$$P(B_m) = P_{m,n} = \underbrace{p^m q^{n-m} + ... + p^m q^{n-m}}_{C_n^m \text{ times}} = C_n^m p^m q^{n-m}.$$

Probabilities $P_{m,n}$ have the form of the terms from the development of the binomial $(p+q)^n$.

This is why the field from this scheme (repartition, distribution) is called the binomial field (its elementary events can be considered elements of the Cartesian product $\Omega^n = \Omega \times ... \times \Omega$).

J. Bernoulli, especially, made this probability scheme the subject of research, and that is why it is also called *Bernoulli scheme.*

The mean and dispersion of a random variable X that is binomially obtained can be easily calculated.

They are $M(X) = np$ and $D^2(X) = npq$.

Example:

Two fighters with equal strength box 12 rounds (the probability for any of them to win a round is 1/2).

Calculate the mean, dispersion and standard deviation of the random variable representing the number of rounds won by one fighter.

Answer:

The random variable X has the binomial repartition:

$$P(X = k) = C_{12}^k \left(\frac{1}{2}\right)^{12}, k = 1, ..., 12.$$

We have $M(X) = 6$, $D^2(X) = 3$ and $D(X) = \sqrt{3}$.

Poisson scheme

This result is also called *the general theorem of repeated experiments*.

Assume n independent experiments are performed, and in each experiment an event A occurs with probability p_i, $(i = 1, 2, ..., n)$ and does not occur with probability $q_i = 1 - p_i$.

We can determine the probability for the event A to occur exactly m times in the n experiments.

Using the same denotations from the binomial scheme, we have:

$$P_{m,n} = p_1 p_2 \cdots p_m q_{m+1} \cdots q_n + ... + p_1 q_2 p_3 \cdots q_{n-1} p_n + ... +$$

$$+ q_1 q_2 \cdots q_{n-m} p_{n-m+1} \cdots p_n .$$

The right-hand member gives the sum of all possible products in which p is shown m times with different indexes, and q is shown $n - m$ times with different indexes.

Considering the products of n binomials $(p_i z + q_i)$, $i = 1, 2, ..., n$,

we have $\varphi_n(z) = \prod\limits_{i=1}^{n} (p_i z + q_i)$, where z is an arbitrary parameter and

$P_{m,n}$ is the coefficient of z^m in this product.

We then have:

$$\prod\limits_{i=1}^{n} (p_i z + q_i) = \sum\limits_{m=0}^{n} P_{m,n} z^m , \text{ with } \sum\limits_{m=0}^{n} P_{m,n} = 1 .$$

Function φ_n is called *the generating function* of probabilities $P_{m,n}$.

Obviously, the particular theorem of repeated experiments (the previous problem) results from the general theorem for $p_1 = p_2 = ... = p_n = p$ and $q_1 = q_2 = ... = q_n = q$.

In this case, the generating function becomes

$$\varphi_n(z) = (pz + q)^n = \sum\limits_{m=0}^{n} C_n^m p^m q^{n-m} z^m .$$

In some applications, the probability for A to occur at least m times in n experiments is required.

Denoting this event by C_m, we have $C_m = B_m \cup B_{m+1} \cup \ldots \cup B_n$; therefore, $P(C_m) = P_{m,n} + P_{m+1,n} + \ldots + P_{n,n}$ or $P(C_m) = \sum_{k=m}^{n} P_{k,n}$.

The probability scheme above described is called *Poisson scheme.*

Example:

At a math contest, three candidates receive one envelope each that contains n ($n > 3$) exam tickets with algebra and geometry problems.

The three envelopes contain 1, 2 and 3 algebra subjects, respectively.

During the test, each of the three candidates randomly extracts a ticket from his or her envelope.

Find the probability of the following events:

1) All three candidates are tested in geometry;
2) None of the candidates is tested in geometry;
3) At least one candidate is tested in algebra.

Answer:

Poisson scheme is applied. We have:

$$\varphi_3(z) = \left(\frac{1}{n}z + \frac{n-1}{n}\right)\left(\frac{2}{n}z + \frac{n-2}{n}\right)\left(\frac{3}{n}z + \frac{n-3}{n}\right) =$$

$$= \frac{1}{n^3}\left[6z^3 + (11n - 18)z^2 + (6n^2 - 22n + 18)z + (n-1)(n-2)(n-3)\right].$$

1) The free term of the polynomial $\varphi_3(z)$ is

$$P_{0,3} = \frac{(n-1)(n-2)(n-3)}{n^3}.$$

2) The coefficient of z^3 from $\varphi_3(z)$ is $\dfrac{6}{n^3}$.

3) $P = 1 - P_{0,3}$.

Assume that in the binomial repartition $P_{m,n} = C_n^m p^m q^{n-m}$ we take $np = \lambda$ (constant).

Let us determine the values of probabilities in this case for $n \to \infty$.

We obtain:

$$\lim_{n\to\infty} P_{k,n} = \lim_{n\to\infty}\left[\frac{n(n-1)...(n-k+1)}{n^k}\frac{\lambda^k}{k!}\left(1-\frac{\lambda}{n}\right)^{n-k}\right] = \frac{\lambda^k}{k!}e^{-\lambda}.$$

Then, by denoting $p_k = \lim_{n\to\infty} P_{k,n}$, we have $p_k = \frac{\lambda^k}{k!}e^{-\lambda}$ and

$$\sum_{k=0}^{\infty} p_k = e^{-\lambda}\sum_{k=0}^{\infty}\frac{\lambda^k}{k!} = e^{-\lambda}\cdot e^{\lambda} = 1.$$

We can see that probabilities p_k are the terms of a repartition.

This repartition is called a *Poisson repartition* of parameter λ, and the random variable

$$X:\begin{pmatrix} 0 & 1 & 2 & ... & k & ... \\ e^{-\lambda} & \frac{\lambda}{1!}e^{-\lambda} & \frac{\lambda^2}{2!}e^{-\lambda} & ... & \frac{\lambda^k}{k!}e^{-\lambda} & ... \end{pmatrix}$$ is called a *Poisson*

random variable.

We can easily calculate $M(X) = \lambda$ and $D^2(X) = \lambda$.

Polynomial scheme

An urn contains balls of colors $c_1, c_2, ..., c_s$ in known proportions, so we know the probability p_i of drawing a ball of color c_i, $i = 1, ..., s$.

n draws of one ball each are performed, provided the urn has the same content after each extraction.

Let A_α be the event α_i *balls of color* $c_i, i = 1, ..., s$ *occur on performed draws,* so $\alpha = (\alpha_1, ..., \alpha_s)$.

The probability of this event is $P(A_\alpha) = \dfrac{n!}{\alpha_1!\alpha_2!...\alpha_s!}p_1^{\alpha_1}...p_s^{\alpha_s}$.

This probability is also denoted by $P(n; \alpha_1, ..., \alpha_s)$.

The described experiment, along with the associated field of events, as probabilized by the formula given above, is called a polynomial scheme and the probability field from this scheme is called a *polynomial field.*

183

Example:

A worker produces a good piece, a piece with a remediable fault and a piece of waste with the probabilities 0.99, 0.07 and 0.03, respectively. The worker produces three pieces.

What is the probability for at least one good piece and at least one waste piece to be among the three pieces produced?

Answer:

We apply the polynomial scheme and find the searched probability to be:

$$P = P(3; 1, 1, 1) + P(3; 2, 0, 1) + P(3; 1, 0, 2) =$$

$$\frac{3!}{1! \cdot 1! \cdot 1!} 0.9 \cdot 0.07 \cdot 0.03 + \frac{3!}{2! \cdot 1!} (0.9)^2 \cdot 0.03 + \frac{3!}{2! \cdot 1!} 0.9 \cdot (0.3)^2 = 0.08667$$

Scheme of nonreturned ball

Let us consider an urn with the following content: a_1 balls of color c_1, a_2 balls of color c_2, ..., a_s balls of color c_s.

n draws are performed without putting back the extracted ball (this experiment is the same as the one in which n balls were drawn once).

Let A_α be the aleatory event *occurrence of exactly α_k balls of color c_k, $k = 1, ..., s$ in the group of the* n *extracted balls,* where

$$\alpha = (\alpha_1, ..., \alpha_s), \quad 0 \le \alpha_k \le a_k, \quad \sum_{k=1}^{s} \alpha_k = n.$$

We have here a field of events that can be probabilized by the classical definition of probability.

The number of equally possible cases is $C_{a_1 + ... + a_s}^n$ (all possibilities of extraction of n balls from the total number of balls from the urn).

The number of cases that are favorable for A_α to occur is $C_{a_1}^{\alpha_1} C_{a_2}^{\alpha_2} ... C_{a_s}^{\alpha_s}$. We then have:

$$P(A_\alpha) = P(n; \alpha_1, ..., \alpha_s) = \frac{C_{a_1}^{\alpha_1} C_{a_2}^{\alpha_2} ... C_{a_s}^{\alpha_s}}{C_{a_1 + ... a_s}^n}.$$

This experiment, along with the associated field of events, probabilized by formula given above, is called a *hypergeometric scheme* or the scheme of nonreturned ball.

Example:
An urn contains thirty-two balls, among which ten balls are white, eight are black, seven are red and five are green.

Five balls are drawn from the urn without putting back any extracted balls.

What is the probability for the following to be among the five drawn balls:

1) three white balls, one red ball and one green ball;
2) one white ball, two black balls and two red balls?

Answer:
We apply the scheme of nonreturned ball. We have:

1) $P(5; 3, 0, 1, 1) = \dfrac{C_{10}^3 C_7^1 C_5^1}{C_{32}^5}$;

2) $P(5; 1, 2, 2, 0) = \dfrac{C_{10}^1 C_8^2 C_7^2}{C_{32}^5}$.

The convergence of sequences of random variables

Let $\{\Omega, \Sigma, P\}$ be a probability σ - field and $X = X(\Omega, \Sigma, P)$ the set of the random variables that are defined on this field.

Call sequence of random variables a function from N to X.

A sequence of random variables is obviously a sequence of functions and is denoted by $(X_n)_{n \in N}$, $(X_n)_{n \geq 0}$ or $(X_n)_{n \geq 1}$.

Definition: The sequence of random variables $(X_n)_{n \geq 1}$ *converges in probability* toward random variable X if for any numbers $\varepsilon, \delta > 0$ a natural number $N(\varepsilon, \delta)$ exists such that $n > N(\varepsilon, \delta)$ implies $P(\{\omega : |X_n(\omega) - X(\omega)| \geq \varepsilon\}) < \delta$.

In this case, we denote this by $X_n \xrightarrow{P} X$.

Proposition

The necessary and sufficient condition for the sequence of random variables $(X_n)_{n \geq 1}$ to converge in probability toward random variable X is that for any $\varepsilon > 0$ a natural number $N(\varepsilon)$ exists such that $n > N(\varepsilon)$ implies $\lim_{n \to \infty} P(\{\omega : |X_n(\omega) - X(\omega)| \geq \varepsilon\}) = 0$.

Proposition

Let $(X_n)_{n \geq 1}$ and $(Y_n)_{n \geq 1}$ be two sequences of random variables. If $X_n \xrightarrow{\;p\;} X$ and $Y_n \xrightarrow{\;p\;} Y$, then $aX_n + bY_n \xrightarrow{\;p\;} aX + bY$, whatever the real value of constants a and b.

Definition: The sequence of random variables $(X_n)_{n \geq 1}$ *converges almost surely* toward random variable X if the set $A = \{\omega : X_n(\omega) \text{ does not tend to } X(\omega)\}$ is an event of Σ and $P(A) = 0$.

In this case, we denote this by $X_n \xrightarrow{\;\text{a.s.}\;} X$.

Proposition

The necessary and sufficient condition for the sequence of random variables $(X_n)_{n \geq 1}$ to converge almost surely toward random variable X is that $\lim_{n \to \infty} P\left(\bigcup_{p=0}^{\infty} \{\omega : |X_{n+p}(\omega) - X(\omega)| \geq \varepsilon\} \right) = 0$ for any $\varepsilon > 0$.

Definition: The sequence of random variables $(X_n)_{n \geq 1}$ *converges in repartition* (or in the Bernoulli sense) toward random variable X if in any continuity point x_0 of distribution function $F(x)$ of X we have $\lim_{n \to \infty} F_n(x_0) = F(x_0)$, where F_n is the distribution function of random variable X_n, $n \in N^*$.

In this case, we denote this by $X_n \xrightarrow{\;r\;} X$.

186

Proposition
If the sequence of random variables $(X_n)_{n\geq 1}$ converges in probability toward X, then it converges in repartition toward X.

Definition: The sequence of random variables $(X_n)_{n\geq 1}$ *converges in* r-*th mean* toward random variable X if exist the absolute moments $M(|X_n|^r)$ (for every $n \in N^*$), $M(|X|^r)$ and if $\lim_{n\to\infty} M(|X_n - X|^r) = 0$.

Proposition
If the sequence of random variables $(X_n)_{n\geq 1}$ converges in r-th mean toward random variable X, then it converges in probability toward X.

The reciprocal is not true.

The next theorem makes the connection between frequency as a random variable and probability:

Law of large numbers

Let $(X_n)_{n\geq 1}$ be a sequence of independent random variables that have the same distribution function (so they have the same mean m and the same dispersion σ^2). Then, for any $\varepsilon > 0$, the sequence

$$\left\{ \omega : \left| \frac{1}{n}\sum_{i=1}^{n} X_i(\omega) - m \right| \geq \varepsilon \right\}$$ converges in probability toward 0 as

$n \to \infty$, that is, $\lim_{n\to\infty} P\left(\left\{ \omega : \left| \frac{1}{n}\sum_{i=1}^{n} X_i(\omega) - m \right| \geq \varepsilon \right\} \right) = 0$.

If we apply the law of large numbers to a sequence $(X_n)_{n\geq 1}$ of independent Bernoullian random variables, then the sum $\frac{1}{n}\sum_{i=1}^{n} X_i(\omega)$ represents the relative frequency $\frac{\alpha}{n}$ of the occurrence of a certain

187

event A, where α is the number of occurrences of event A in n independent successive experiments.

According to this theorem, we deduce that $\lim\limits_{n\to\infty} P\left(\left|\dfrac{\alpha}{n} - p\right| \geq \varepsilon\right) = 0$.

This classical result is known as *Bernoulli's theorem*:

The relative frequency converges in probability toward probability (in this enunciation, the word *probability* appears in two different mathematical senses).

A generalization of Bernoulli's theorem is the following:

Poisson's Theorem

If α is the number of occurrences of an event A in n independent experiments, and p_k is the probability of the occurrence of event A in the k-th experiment, $k \in N^*$, then the limit $\lim\limits_{n\to\infty} \dfrac{p_1 + \ldots + p_n}{n} = p$

exists and $\left(\dfrac{\alpha}{n}\right)_{n\in N^*}$ converges in probability toward p.

The law of large numbers reveals the existence of sequences of random variables that tend in probability toward a constant.

This law is valid for an ensemble of phenomena and not for each separate phenomenon.

Also, it does not stand for a property of any sequence of random variables.

This chapter contains the minimal mathematics necessary to understand the probability concept in the sense of its mathematical definition and to solve practical applications for finite and discrete fields of events.

We presented here all the basic notions of probability theory and the introductory notions of mathematical statistics, along with the theoretical results that make an important contribution in solving practical applications.

Only the discrete cases were presented (discrete probability fields, discrete random variables and particularizations of the main theorems for the discrete case).

The readers who want more knowledge of this domain may choose to consult more complex papers on probability theory and mathematical statistics.

Using the theoretical results presented here, together with the combinatorics calculus formulas from the next chapter, any reader with a minimal mathematical background should have all the tools necessary to perform probability calculations.

How easily the reader performs these calculations, as well as their accuracy, is only a matter of training and practice.

COMBINATORICS

Combinatorial analysis plays a major role in probability applications, from a calculus perspective, because many situations deal with permutations, combinations or arrangements.

The correct approach to combinatorial problems and the ease of handling combinatorial calculus are 50 percent of the probability calculus abilities for games of chance.

Therefore, this chapter contains many solved and unsolved practical applications that are useful for learning this calculus.

As in the previous chapter on mathematics, the theoretical discussions present only definitions and important results, without demonstrations.

Permutations

Definition: Call the *permutations* of a set A with n elements all the ordered sets that can be generated by the n elements of A.

By ordered set we mean a set associated with the order of its elements (two ordered sets in which the same elements are differently enumerated are considered different).

An ordered set obtained through a rearrangement of elements is called a permutation.

Example:
The permutations of the set $\{a, b, c\}$ are (abc), (bca), (cab), (acb), (bac) and (cba).

Permutation formula
The number of permutations of n elements ($n \in N^*$) is:
$P_n = 1 \cdot 2 \cdot 3 \cdot \ldots \cdot n = n!$ (factorial of n); $0! = 1$ (by convention)

Examples:

1) The number of permutations of three elements a, b and c (abc, bca, cab, acb, bac, cba) is $P_3 = 1 \cdot 2 \cdot 3 = 3! = 6$.

2) The number of ways in which we can arrange five books on a shelf is $5! = 1 \cdot 2 \cdot 3 \cdot 4 \cdot 5 = 120$.

Arrangements

Definition: Call the *arrangements* of n elements taken m at a time ($m \leq n$) of a set A with n elements, all the ordered subsets with m elements that can be generated by the n elements of A.

In fact, the arrangements are partial permutations of the elements of the initial set.

Example:
The arrangements of the set $\{a, b, c\}$ taken two at a time are:
(ab), (ba), (ac), (ca), (bc) and (cb).

Arrangement formula
The number of arrangements of n elements taken m at a time is:
$$A_n^m = n(n-1)(n-2)...(n-m+1) = \frac{n!}{(n-m)!}, \quad m \leq n.$$
For this number, the denotations *A(n, m)* or *nPm* are also used.

Examples:

1) The number of arrangements of three elements a, b, c taken two at a time (ab, ac, bc, ba, ca, cb) is $A_3^2 = 3 \cdot 2 = 6$.
2) At an athletic race with five competitors, the number of possible rankings for the first three places is $A_5^3 = 5 \cdot 4 \cdot 3 = 60$.

Properties
$A_n^n = P_n = n!$, $A_n^{n-1} = A_n^n$, $A_n^0 = 1$.

Combinations

Definition: We name combinations of n elements taken m at a time ($m \leq n$) of a set A with n elements all the subsets with m elements that can be generated by the n elements of A.

Observe that this definition differs from the definition of arrangements by the omission of the word *ordered.*
The order of elements does not count toward combinations.
Their definition reverts to the classical definition of subsets; in fact, a combination is a subset (in which the order of elements does not count).
The combination (abc) is identical to (acb) or (cba).
A combination of n elements taken m at a time is also called a m-size combination from an n-element set.

Example:
The combinations of the set $\{a, b, c\}$ taken two at a time are (ab), (ac) and (bc). The only 3-size combination of this set is (abc).

Combination formula
The number of combinations of n elements taken m at a time is:

$$C_n^m = \frac{A_n^m}{m!} = \frac{n!}{m!(n-m)!} = \frac{n(n-1)...(n-m+1)}{m!}.$$

The denotations $C(n, m)$, nCm or $\binom{m}{n}$ are also used for this

number.

Examples:

1) The number of combinations of three elements a, b, c taken two at a time (ab, ac, bc) is $C_3^2 = \dfrac{3 \cdot 2}{2} = 3$.

2) If we randomly choose two cards from a 52-card deck, the number of all possible choices is $C_{52}^2 = \dfrac{51 \cdot 52}{2} = 1326$.

Properties

$$C_n^1 = n, \quad C_n^n = C_n^0 = C_0^0 = 1.$$

$$C_n^m = C_n^{n-m}, \quad C_n^m = C_{n-1}^m + C_{n-1}^{m-1}.$$

$$C_n^{k+1} = \frac{n-k}{k+1} C_n^k.$$

$$\sum_{k=0}^{n} C_n^k = 2^n$$ (this is also the number of all subsets of an *n*-element set).

Combinatorial analysis has several applications in mathematical branches such as algebra, trigonometry, and real and complex analysis.

It is an essential tool for the various applications of probability calculus, especially in games of chance, when counting the possible and favorable variants or gaming situations, with respect to the occurrence of a certain event.

Combinatorial calculus

In combinatorics, calculating combinations simply means counting.

Any calculus problem will ask for the number of certain combinations (arrangements, permutations).

In case there are no additional hypotheses on certain elements of the combinations we deal with, this number can be obtained by direct application of the corresponding formula.

If the problem comes up with additional conditions on the elements of combinations, the solution will also use properties of combinations before applying the formula directly.

For example, the problem "How many ways can we arrange the letters a, b, c, d, e in 4-size groups?" requires an immediate application of arrangements' formula of A_n^m, by replacing n and m with the numerical values n = 5 and m = 4.

But the problem "How many ways can we arrange the letters a, b, c, d, e in 4-size groups, such that each group begins with a vowel?" requires an extra analysis, direct application of formula being no longer possible.

One of the most commonly used methods of solution involving the properties of combinations is partitioning after the calculus formulas are applied to smaller combinations.

We explain this procedure in detail later in this chapter.

Because the large majority of probability calculus applications using combinatorics deal solely with combinations, we insist on using these combinations in the calculus methodology and applications presented here.

Direct application of formulas

Often, we can directly apply the calculus formulas presented at the beginning of this chapter by replacing the variables with the parameters of each problem.

This direct application can be done in the case of problems having no additional conditions on certain elements of the combinations, which means their elements keep the same generality (see the example above).

After replacing the parameters in the formula, the numerical calculus can take place, which involves only operations of multiplication and division (reduction of fractions).

Acquiring the skill to perform factorial calculus assumes practice and memorization as well.

Memorizing some factorial products is useful to avoid repeated multiplications and to insert precalculated values in complicated expressions.

The following factorial products can be easily memorized:

$2! = 2$; $3! = 6$; $4! = 24$; $5! = 120$; $6! = 720$; $7! = 5040$; $8! = 40320$.

For arrangements and combinations, calculus formulas can be applied in either of two forms (factorial or developed).

The developed form is obtained from the factorial one by enumerating the factors and reducing them in the fraction:

$$A_n^m = \frac{n!}{(n-m)!} = \frac{1 \cdot 2 \cdot ... \cdot (n-m) \cdot (n-m+1) \cdot (n-m+2) \cdot ... \cdot n}{1 \cdot 2 \cdot ... \cdot (n-m)} =$$
$$= (n-m+1)(n-m+2)...(n-1)n.$$

$$C_n^m = \frac{n!}{m!(n-m)!} = \frac{1 \cdot 2 \cdot ... \cdot (n-m) \cdot (n-m+1) \cdot (n-m+2) \cdot ... \cdot n}{m!(1 \cdot 2 \cdot ... \cdot (n-m))} =$$
$$= \frac{(n-m+1) \cdot (n-m+2) \cdot ... \cdot n}{1 \cdot 2 \cdot ... \cdot m}.$$

In detail, the minimal numerical calculus for the two formulas is represented by the following algorithm:

a) **For arrangements** A_n^m :
1) Calculate the difference $n - m$.
2) Calculate the product of all consecutive numbers from $n - m + 1$ to n.

Examples:

– Let us calculate A_7^3 :
1) $7 - 3 = 4$
2) Calculate the product of all numbers from $4 + 1 = 5$ to 7; namely, $5 \cdot 6 \cdot 7 = 210$.
$$A_7^3 = 210$$

– Let us calculate A_{10}^5 :
1) $10 - 5 = 5$
2) Calculate the product of all numbers from $5 + 1 = 6$ to 10; namely, $6 \cdot 7 \cdot 8 \cdot 9 \cdot 10 = 42 \cdot 720 = 3024$.
$$A_{10}^5 = 3024$$

Other examples:
$$A_8^3 = 6 \cdot 7 \cdot 8 = 42 \cdot 8 = 336$$
$$A_{17}^7 = 11 \cdot 12 \cdot 13 \cdot 14 \cdot 15 \cdot 16 \cdot 17 = 98017920$$

b) **For combinations** C_n^m :
1) Calculate the difference $n - m$.
2) Calculate the product of all consecutive numbers from $n - m + 1$ to n.
3) Calculate the product of all consecutive numbers from 1 to m.
4) Divide the result obtained in step 2 by the result obtained in step 3.

197

Examples:

– Let us calculate C_5^2 :

1) $5 - 2 = 3$

2) Calculate the product of all numbers from $3 + 1 = 4$ to 5; namely, $4 \cdot 5 = 20$.

3) Calculate the product of numbers from 1 to 2; namely, $1 \cdot 2 = 2$.

4) Divide the result obtained in step 2 by the result obtained in step 3, namely $20 : 2 = 10$.

$C_5^2 = 10$

– Let us calculate C_{12}^5 :

1) $12 - 5 = 7$

2) Calculate the product of all numbers from $7 + 1 = 8$ to 12; namely, $8 \cdot 9 \cdot 10 \cdot 11 \cdot 12$.

3) Calculate the product of all numbers from 1 to 5; namely, $1 \cdot 2 \cdot 3 \cdot 4 \cdot 5$.

4) Divide the result from step 2 by the result from step 3, by writing it as the fraction $\dfrac{8 \cdot 9 \cdot 10 \cdot 11 \cdot 12}{1 \cdot 2 \cdot 3 \cdot 4 \cdot 5}$; after the immediate reductions, we get $8 \cdot 9 \cdot 11 = 792$.

$C_{12}^5 = 792$

Leaving the products from steps 2 and 3 in their uncalculated factorial form is recommended, because this allows the reduction of the fraction from step 4, and this operation spares us of a lot of additional calculations.

Other examples:

$$C_{15}^3 = \frac{13 \cdot 14 \cdot 15}{1 \cdot 2 \cdot 3} = 13 \cdot 7 \cdot 5 = 13 \cdot 35 = 455$$

$$C_{19}^4 = \frac{16 \cdot 17 \cdot 18 \cdot 19}{1 \cdot 2 \cdot 3 \cdot 4} = 4 \cdot 17 \cdot 3 \cdot 19 = 12 \cdot 17 \cdot 19 = 3876$$

The property $C_n^m = C_n^{n-m}$ is useful in calculations, in the sense of simplifying them.

This property of combinations must be used when the difference $n - m$ is less than m, because it reduces the number of product factors.

For example, for n = 57 and m = 53, we have $C_{57}^{53} = C_{57}^4$.

Obviously, C_{57}^4 is more easily developed and calculated by applying the formula than C_{57}^{53} because it contains the factorial 4! (instead of 53!):

$$C_{57}^4 = \frac{54 \cdot 55 \cdot 56 \cdot 57}{4!}, \quad C_{57}^{53} = \frac{5 \cdot 6 \cdot \ldots \cdot 57}{53!}.$$

The property $C_n^m = C_n^{n-m}$ must be applied, in general, when m and n are close in value to one another.

Examples:

$$C_7^5 = C_7^2 = \frac{6 \cdot 7}{2} = 21$$

$$C_{11}^8 = C_{11}^3 = \frac{9 \cdot 10 \cdot 11}{2 \cdot 3} = 3 \cdot 5 \cdot 11 = 165$$

$$C_{25}^{19} = C_{25}^6 = \frac{20 \cdot 21 \cdot 22 \cdot 23 \cdot 24 \cdot 25}{2 \cdot 3 \cdot 4 \cdot 5 \cdot 6} = 10 \cdot 7 \cdot 22 \cdot 23 \cdot 5 = 177100$$

At the final step of the calculus algorithm, after we obtain the last fraction, we can check for possible errors by following both the numerator and the denominator, which must show all the consecutive factors of the products from previous steps.

Because the number of combinations (arrangements, permutations) is natural, total reduction of that fraction (complete evanescence of the denominator) is obligatory.

If we find that the fraction cannot be reduced in totality, it is a sign that an error has occurred at the previous steps, and that calculus must be redone.

Partitioning combinations

Let us do the following exercise: unfold all 4-size combinations from the numbers (1, 2, 3, 4, 5, 6, 7).

The whole number of these must be $C_7^4 = C_7^3 = 35$.

This unfoldment must not be aleatory because doing otherwise risks losing combinations; in addition, some combinations end up being written more than once.

To make a proper unfoldment, we use the following algorithm:

While permanently maintaining the ascending order of the numbers from written combinations (from left to right), we successively increase the numbers starting from the right.

When increasing the numbers is no longer possible, we increase the number from the first previous place of combination and continue the procedure until the capacity to increase is no longer possible.

Choosing the ascending order does not affect the overall approach to solving the problem because the order of elements does not count within a combination.

This procedure ensures that all combinations are enumerated, without omission, and avoids any repetition (double counting).

Here is how this unfoldment works concretely for the previously proposed problem:

Start with the combination (1234). Successively increase number 4 from the last (fourth) place: (1235), (1236), (1237).

Increasing the number from the last place is no longer possible.

Increase number 3 from the third place by replacing it with 4: (1245), (1246), (1247).

Follows to replace 4 with 5 on the third place: (1256), (1257).

Replace 5 with 6 on the third place: (1267).

Follows to change the second place: replace 2 with 3 and start with 4 on the third place: (1345), (1346), (1347).

Replace 4 with 5 on the third place: (1356), (1357).

Replace 5 with 6 on the third place: (1367).

Again follows to change the second place with 4 instead of 3: (1456), (1457).

Replace 5 with 6 on the third place: (1467).

Replace 4 with 5 on the second place: (1567).

Now, the combinations having number 1 on the first place are exhausted. Replace 1 with 2 on the first place and start with 3 on the second place and 4 on the third place: (2345), (2346), (2347).

Put 5 on the third place: (2356), (2357).

Put 6 on the third place: (2367).

Replace 3 with 4 on the second place, and start with 5 on the third place: (2456), (2457).

Put 6 on the third place: (2467).

Replace 4 with 5 on the second place: (2567).

Now, the combinations having number 2 on the first place are exhausted. Put 3 on the first place and start with 4 on the second place and 5 on the third place: (3456), (3457).

Put 6 on the third place: (3467).

Replace 4 with 5 on the second place: (3567).

Now, the combinations having number 3 on the first place are also exhausted. Replace 3 with 4 on the first place: (4567).

This is the last counted combination.

By counting all these unfolded combinations, we find thirty-five different combinations; therefore, the unfoldment is correct.

We performed this algorithmic unfoldment in order to grasp the effective process of generating the combinations of elements of a given set and to see why repeated multiplications are the only elementary operations involved in the process of counting.

In the example above, observe that:

– For counting the 4-size combinations containing number 1, we fix this number and unfold all 3-size combinations containing the rest of numbers (2, 3, 4, 5, 6, 7), which are in number of $C_6^3 = 20$.

Indeed, by effectively counting the unfolded combinations containing 1, we find twenty combinations.

A combination containing the number 1 can be written as (1xyz), where x, y and z are mutually different variables with values in the set $\{2, 3, 4, 5, 6, 7\}$.

By an abuse of denotation, we can write (1xyz) = (1)(xyz) = $1 \cdot C_6^3 = 20$, this means: the number of 4-size combinations containing the number 1 is equal to number of 1-size combinations containing 1 (namely, a single combination) multiplied by the number of 3-size combinations from the set of numbers left (2, 3, 4, 5, 6, 7).

Obviously, the same result (twenty) also stands for the number of combinations containing any of the numbers 2, 3, 4, 5, 6 or 7.

Note that the total number of 4-size combinations of elements from the given set is not equal to the sum of these partial results (number of combinations containing 1 + number of combinations containing 2 + ... + number of combinations containing 7)!

This can be verified immediately ($35 \neq 20 \cdot 6$) and is explained by the fact that, through addition, some combinations are counted more than once (a combination containing 1 could also contain 2, etc.).

– For counting the 4-size combinations containing the numbers 1 and 2, we fix the numbers 1 and 2 and unfold all 2-size combinations of elements from the set of numbers left (3, 4, 5, 6, 7), in number of $C_5^2 = 10$.

By the same abuse of denotation, we can write $(12xy) = (12)(xy)$ $= 1 \cdot C_5^2 = 10$: the number of 4-size combinations containing the numbers 1 and 2 is equal to number of 2-size combinations containing 1 and 2 (namely, a single combination) multiplied by the number of 2-size combinations from the set of numbers left (3, 4, 5, 6, 7).

Obviously, the same result (ten) also stands for the number of combinations containing any two given numbers (23), (35), (57), etc.

– For counting the 4-size combinations containing the numbers 1, 2 and 3, we fix the numbers 1, 2 and 3 and unfold all 1-size combinations of elements from the set of numbers left (4, 5, 6, 7), in number of $C_4^1 = 4$.

By the same abuse of denotation, we can write $(123x) = (123)(x)$ $= 1 \cdot 4 = 4$: the number of 4-size combinations containing the numbers 1, 2 and 3 is equal to number of 3-size combinations containing 1, 2 and 3 (namely, a single combination) multiplied by the number of 1-size combinations from the set of numbers left (3, 4, 5, 6, 7).

Obviously, the same result (four) also stands for the number of combinations containing any three given numbers (234), (357), (136), etc.

In the previous exercise, the count of combinations containing given numbers having the form (1xyz), (12xy) or (123x) stands for classic examples of problems that impose additional conditions on the elements of combinations.

Their solution uses the graphic representations (1xyz) = (1)(xyz), (12xy) = (12)(xy) and (123x) = (123)(x), which simplify the calculus by reductions to lower size combinations.

Call this procedure *partitioning* of combinations.

The immediate generalization of this exercise is as follows:

Let $E = A \cup B \cup C \cup D$ be a set, with A, B, C and D being mutually exclusive.

If from all 4-size combinations of elements of E we want to count those containing one element from A, we write:

(abcb) = (a)(bcd), where $a \in A$ (this is the additional condition imposed on the elements of combinations); (a) is the number of elements of A and (bcd) the number of 3-size combinations of elements from $B \cup C \cup D$; these two numbers are multiplied; observe that the sets A and $B \cup C \cup D$ are disjoint.

If we want to count the combinations containing two elements from $A \cup B$, we write:

(abcd) = (ab)(cd), where $a, b \in A \cup B$; (ab) is the number of 2-size combinations of elements of $A \cup B$ and (cd) the number of 2-size combinations of elements of $C \cup D$; observe that the sets $A \cup B$ and $C \cup D$ are exclusive (this is an obligatory condition for the partitioning procedure to be applied; otherwise, some combinations would be counted more than once).

The procedure stands for combinations of any size, regardless of the conditions on their elements.

The generalization is as follows:

Let us consider the n-size combinations $(a_1 a_2 ... a_n)$ of elements of a finite set A. Let $(k_1, k_2, ..., k_m)$ be a partition of n (k_i are natural numbers such that $k_1 + k_2 + ... + k_m = n$) and $A_{k_1}, A_{k_2}, ..., A_{k_m}$ a partition of the set A (A_{k_i} are mutually exclusive sets such that

$\bigcup\limits_{i=1}^{m} A_{k_i} = A$). Then:

$$(a_1 a_2 ... a_n) = (a_1 a_2 ... a_{k_1})(a_{k_1+1} ... a_{k_1+k_2})...(a_{k_1+...+k_{m-1}+1} ... a_n)$$

As we stated earlier, this is an abuse of denotation and represents a procedure rather than a formula.

The partitioning of combinations is, in fact, a graphic representation that allows us to view a property and simplify the calculus.

Graphic representations, and literal denotations, even those that are considered abusive, are highly recommended in combinatorial problems. They aid the correct framing of a problem by applying the proper properties and the correct performance of calculus.

The following examples show how these procedures work in concrete applications.

Applications

Solved applications

1) We have fifteen books and we must fill a shelf that can accommodate only eleven books.

a) How many ways can we arrange eleven books in the shelf, by choosing from the fifteen?

b) How many ways can we choose the eleven books?

Solution:

a) We can directly apply the arrangement formula; the searched number is A_{15}^{11}; according to the minimal calculus algorithm for arrangements, we find:

1) $15 - 11 = 4$

2) Do the product of numbers from $4 + 1 = 5$ to 15:

$5 \cdot 6 \cdot 7 \cdot 8 \cdot 9 \cdot 10 \cdot 11 \cdot 12 \cdot 13 \cdot 14 \cdot 15 =$

$= 30 \cdot 56 \cdot 90 \cdot 132 \cdot 13 \cdot 14 \cdot 15 = 54486432000$.

$A_{15}^{11} = 54486432000$

b) The choices do not take order into account.

We deal here only with combinations, and the searched number is C_{15}^{11}.

We may use the property $C_{15}^{11} = C_{15}^{15-11} = C_{15}^{4}$ to reduce the calculus and then follow the minimal calculus algorithm for combinations:

1) $15 - 4 = 11$

2) Do the product of numbers from 12 to 15: $12 \cdot 13 \cdot 14 \cdot 15$

3) Do the product of numbers from 1 to 4: $2 \cdot 3 \cdot 4$

4) Do the ratio $\dfrac{12 \cdot 13 \cdot 14 \cdot 15}{2 \cdot 3 \cdot 4}$; after reductions, we get

$13 \cdot 7 \cdot 15 = 1365$.

$C_{15}^{11} = 1365$

2) How many ways can we arrange the letters of the word MAJORITY?

Solution:
We deal here with permutations. The letters are different and total seven, so the total number of permutations is:
$$7! = 2 \cdot 3 \cdot 4 \cdot 5 \cdot 6 \cdot 7 = 24 \cdot 30 \cdot 7 = 24 \cdot 210 = 5040.$$

3) At 6/49 lottery system (six numbers are drawn from a total of forty-nine, from 1 to 49, with one played variant having six numbers), calculate the total number of possible variants that can be drawn.

Solution:
A drawn variant represents a 6-size combination from forty-nine numbers, so the searched number is C_{49}^6 :

1) $49 - 6 = 43$
2) Do the product $44 \cdot 45 \cdot 46 \cdot 47 \cdot 48 \cdot 49$
3) Do the product $2 \cdot 3 \cdot 4 \cdot 5 \cdot 6$
4) Do the ratio
$$\frac{44 \cdot 45 \cdot 46 \cdot 47 \cdot 48 \cdot 49}{2 \cdot 3 \cdot 4 \cdot 5 \cdot 6} = 44 \cdot 3 \cdot 46 \cdot 47 \cdot 48 \cdot 49 = 13983816.$$
$$C_{49}^6 = 13983816$$

(Let us hope that gamblers who usually play using a few variants are not too disappointed!)

4) In a 6/49 lottery system, calculate how many possible variants containing the numbers 5 and 11 exist.

Solution:
By fixing the two numbers, the variants containing them will have the form (5 11 xyzt), where x, y, z and t are distinct numbers belonging to the set $\{1, 2, ..., 49\} - \{5, 11\}$, which has $49 - 2 = 47$ elements.
We do the partitioning $(5\ 11\ xyzt) = (5\ 11)(xyzt)$.

(5 11) represents one combination and the number of (xyzt) combinations is given by the number of 4-size sets that can be built from the numbers left (47), namely, C_{47}^4.

Then, the searched number is

$$1 \cdot C_{47}^4 = C_{47}^4 = \frac{44 \cdot 45 \cdot 46 \cdot 47}{2 \cdot 3 \cdot 4} = 178365.$$

5) In a 6/49 lottery system, calculate how many variants containing the numbers 1, 2 and 3 exist.

Solution:
The respective variants will have a (123xyz) form, with x, y and z mutually different and different from 1, 2 and 3. The set from which x, y and z can take values has 46 elements (49 – 3).

By partitioning, we have (123xyz) = (123)(xyz).
Denoting by C the searched number, we have:

$$C = 1 \cdot C_{46}^3 = C_{46}^3 = \frac{44 \cdot 45 \cdot 46}{2 \cdot 3} = 15180 \quad (1 \text{ is the number of (123)}$$

combinations and C_{46}^3 is the number of (xyz) combinations).

6) In a 6/49 lottery system, calculate: a) how many variants containing only even numbers exist; b) how many variants containing only uneven numbers exist.

Solution:
a) The even numbers are 2, 4, 6, 8, ..., 48, and number 24 in total (we have $2 = 1 \cdot 2$, $4 = 2 \cdot 2$, $6 = 3 \cdot 2$, ..., $48 = 24 \cdot 2$; the count was done by following the first factor of the products).

The number of combinations of six even numbers is then

$$C_{24}^6 = \frac{19 \cdot 20 \cdot 21 \cdot 22 \cdot 23 \cdot 24}{2 \cdot 3 \cdot 4 \cdot 5 \cdot 6} = 134596.$$

b) The uneven numbers are 1, 3, 5, 7, ..., 49, and total 49–24=25 (we have subtracted the number of even numbers). The number of combinations of six uneven numbers is then

$$C_{25}^6 = \frac{20 \cdot 21 \cdot 22 \cdot 23 \cdot 24 \cdot 25}{2 \cdot 3 \cdot 4 \cdot 5 \cdot 6} = 177100.$$

7) In a 6/49 lottery system, find the number of possible variants containing exactly four even numbers.

Solution:
Let us denote such variant by (PPPPxy), where P are less than forty-nine even numbers and x, y are distinct and uneven.

This is obviously an abuse of denotation. Repeating the same letter P does not mean those numbers are equal (a combination cannot contain identical elements); it only means they are even and mutually different.

This denotation simplifies the partitioning and calculus by showing the elements on which the additional conditions were imposed.

(PPPPxy) = (PPPP)(xy)

The number of (PPPP) combinations is the number of 4-size combinations from twenty-four (the number of even numbers); namely, C_{24}^4, and the number of (xy) combinations is the number of 2-size combinations from twenty-five (the rest of numbers; namely, the uneven ones), respectively C_{25}^2.

We then calculate:

$$C_{24}^4 \cdot C_{25}^2 = \frac{21 \cdot 22 \cdot 23 \cdot 24}{2 \cdot 3 \cdot 4} \cdot \frac{24 \cdot 25}{2} = 3187800.$$

8) In a 5/40 lottery system (six numbers are drawn from a total of forty, from 1 to 40, with one played variant having five numbers), calculate the total number of possible variants that can be drawn.

Solution:
The number is given by the number of 6-size sets that can be built from the forty numbers; namely, $C_{40}^6 = 3838380$.

9) In a 5/40 lottery system, find the number of possible variants played that contain exactly two numbers higher than 17.

Solution:
A played variant contains five numbers. Let us denote by (NNxyz) a variant fitting the problem's conditions, where $N > 17$ and x, y and z are mutually different and take values in the set of

numbers less or equal to 17 (so, obviously, they are also different from the first two elements of the combination).

(NNxyz) = (NN)(xyz)

The set from which N takes values has $40 - 17 = 23$ elements and the set from which x, y and z take values has seventeen elements.

The searched number is then

$$C_{23}^2 \cdot C_{17}^3 = \frac{22 \cdot 23}{2} \cdot \frac{15 \cdot 16 \cdot 17}{2 \cdot 3} = 172040 \,.$$

10) In a 5/40 lottery system, find the number of drawn variants possible that contain at least three numbers larger than 21.

Solution:

A drawn variant has six numbers. The respective combinations will have a (NNNxyz) form, with $N > 21$ (by an abuse of denotation, NNN represents distinct elements larger than 21) and any x, y and z (they could take values even larger than 21).

At this stage, we cannot do the partitioning (NNNxyz) = (NNN)(xyz), because the two attached sets of values are not exclusive (x, y and z can take values from the first set; namely, the one of numbers larger than 21).

Such partitioning, followed by multiplication, is a frequent error. Let us unfold this incorrect calculus to its end:

The number of combinations (NNN) is C_{19}^3, because the numbers larger than 21 are 22, 23, ..., 40, and number 19 in total.

The number of combinations (xyz) is $C_{40-3}^3 = C_{37}^3$ (for each combination (NNN), remain $40 - 3$ numbers from which we unfold the 3-size combinations).

The searched number should then be $C_{19}^3 \cdot C_{37}^3$.

As we said, this calculus procedure is wrong because the condition has not been met for the sets of values attached to that partitioning to be exclusive. The number becomes larger than the real number because some combinations are counted several times.

For example, as a result of wrong partitioning, the following combinations:

(25 22 35)(23 5 17)
(25 23 22)(35 5 17)
N N N   x y z

were counted as distinct, while in fact they are one and same combination.

The correct calculus algorithm consists of splitting the problem into several partial problems: we calculate the number of combinations containing exactly three N-numbers (larger than 21), four N-numbers, five N-numbers, and six N-numbers, and then add the results.

1) Exactly three N-numbers

The searched combinations have a (NNNxyz) form, with $x,y,z \neq N$ ($x, y, z \leq 21$). The condition for x, y and z to be mutually distinct is implied and is not written from here on to simplify the text.

(NNNxyz) = (NNN)(xyz)

The N-numbers may take nineteen values (from 22 to 40) and x, y, z may take twenty-one values (from 1 to 21). Obviously, the two sets of values are exclusive.

The searched number is then

$$C_{19}^3 \cdot C_{21}^3 = \frac{17 \cdot 18 \cdot 19}{2 \cdot 3} \cdot \frac{20 \cdot 21 \cdot 22}{2 \cdot 3} = 17 \cdot 3 \cdot 19 \cdot 10 \cdot 7 \cdot 22 = 1492260.$$

2) Exactly four N-numbers

The combinations have a (NNNNxy) form, with $x,y \neq N$.

(NNNNxy) = (NNNN)(xy)

N may take nineteen values, while x and y may take twenty-one values.

The number of combinations is

$$C_{19}^4 \cdot C_{21}^2 = \frac{16 \cdot 17 \cdot 18 \cdot 19}{2 \cdot 3 \cdot 4} \cdot \frac{20 \cdot 21}{2} = 2 \cdot 17 \cdot 6 \cdot 19 \cdot 10 \cdot 21 = 813960.$$

3) Exactly five N-numbers

The combinations have a (NNNNNx) form, with $x \neq N$.

(NNNNNx) = (NNNNN)(x)

N may take nineteen values and x may take twenty-one values.

The number of combinations is

$$C_{19}^5 \cdot C_{21}^1 = \frac{15 \cdot 16 \cdot 17 \cdot 18 \cdot 19}{2 \cdot 3 \cdot 4 \cdot 5} \cdot 21 = 2 \cdot 17 \cdot 18 \cdot 19 \cdot 21 = 244188.$$

4) Six N-numbers

The combinations have a (NNNNNN) form, in number of

$$C_{19}^6 = \frac{14 \cdot 15 \cdot 16 \cdot 17 \cdot 18 \cdot 19}{2 \cdot 3 \cdot 4 \cdot 5 \cdot 6} = 14 \cdot 3 \cdot 2 \cdot 17 \cdot 19 = 27132 \,.$$

Now we add the partial results from 1), 2), 3) and 4), being sure no combinations are counted more than once:

1492260 + 813960 + 244188 + 27132 = 2577540

The number of drawn combinations possible that contain at least three numbers larger than 21 is 2577540.

This partial calculus procedure, followed by adding, is applied in every similar situation (problems in which the condition imposed on the elements of combinations is of the *at least...* type).

Through this procedure, any duplicate counting can be avoided.

11) At a 52-card classical poker game (in which each player is dealt five cards initially, from which he or she holds some cards and replaces the rest with new ones in the second distribution), find the total number of possible hands a player can be dealt at the first card distribution (how many ways five cards can be dealt to a player).

Solution:

The player receives an arbitrary five-card combination from 52, so the total number of possible combinations is

$$C_{52}^5 = \frac{48 \cdot 49 \cdot 50 \cdot 51 \cdot 52}{2 \cdot 3 \cdot 4 \cdot 5} = 2 \cdot 49 \cdot 10 \cdot 51 \cdot 52 = 2598960 \,.$$

12) From all combinations found in the previous exercise, find how many of them contain:

a) exactly two clubs;

b) a minimum of two clubs;

c) five cards with the same symbol (suit) (clubs ♣, spades ♠, diamonds ♦, hearts ♥); and

d) a pair of queens (QQ).

Solution:
a) Denote the clubs card by *C*.

A combination containing exactly two clubs has a (CCxyz) form, with x, y, z ≠ C.

(CCxyz) = (CC)(xyz); *C* can take thirteen values (there are thirteen clubs cards); *x*, *y* and *z* can take 52 − 13 = 39 values.

The searched number is then

$$C_{13}^2 \cdot C_{39}^3 = \frac{12 \cdot 13}{2} \cdot \frac{37 \cdot 38 \cdot 39}{2 \cdot 3} = 6 \cdot 13 \cdot 37 \cdot 37 \cdot 19 \cdot 13 = 712842 .$$

b) We successively find the number of combinations containing exactly three, four and five clubs and then add them together with the number found at a):

1) Exactly three clubs

(CCCxy) = (CCC)(xy); *C* takes thirteen values, while *x* and *y* take thirty-nine values. The number of combinations is

$$C_{13}^3 \cdot C_{39}^2 = \frac{11 \cdot 12 \cdot 13}{2 \cdot 3} \cdot \frac{38 \cdot 39}{2} = 11 \cdot 2 \cdot 13 \cdot 19 \cdot 39 = 211926 .$$

2) Exactly four clubs

(CCCCx) = (CCCC)(x); *C* takes thirteen values and *x* takes thirty-nine values. The number of combinations is

$$C_{13}^4 \cdot 39 = \frac{10 \cdot 11 \cdot 12 \cdot 13}{2 \cdot 3 \cdot 4} \cdot 39 = 5 \cdot 11 \cdot 13 \cdot 39 = 27885 .$$

3) Five clubs

The number of (CCCCC) combinations is

$$C_{13}^5 = \frac{9 \cdot 10 \cdot 11 \cdot 12 \cdot 13}{2 \cdot 3 \cdot 4 \cdot 5} = 9 \cdot 11 \cdot 13 = 1287 .$$

In total (together with the result from a)), we have
712842 + 211926 + 27885 + 1287 = 953940 combinations containing at least two clubs.

c) For a specific symbol (let *S* be the symbol), the number of (SSSSS) combinations is $C_{13}^5 = 1287$ (see b)1)).

We have four symbols (♣, ♠, ♦, ♥); therefore, the searched number is $4 \cdot C_{13}^5 = 4 \cdot 1287 = 5148$.

d) The phrase *a pair of queens* can stand for *exactly two queens* or *at least two queens.* We will do the calculation for each instance.

1) Exactly two queens

The combinations have a (QQxyz) form, with $x, y, z \neq Q$; Q takes four values (Q♣, Q♠, Q♦, Q♥); x, y and z take $52 - 4 = 48$ values. Their number is

$$C_4^2 \cdot C_{48}^3 = \frac{3 \cdot 4}{2} \cdot \frac{46 \cdot 47 \cdot 48}{2 \cdot 3} = 6 \cdot 46 \cdot 47 \cdot 8 = 103776.$$

2) Exactly three queens

The combinations have a (QQQxy) form, with $x, y \neq Q$.

Their number is $C_4^3 \cdot C_{48}^2 = 4 \cdot \frac{47 \cdot 48}{2} = 4512$.

3) Four queens

The combinations have a (QQQQx) form, with $x \neq Q$, which number $C_4^4 \cdot 48 = 1 \cdot 48 = 48$.

The number of combinations containing at least two queens is then $103776 + 4512 + 48 = 108336$.

13) Same problem for a 32-card classical poker game (*to be solved*).

14) In Texas Hold'em Poker (in which each player is first dealt two cards (called pocket or hole cards) from a deck of 52; then the dealer places five common cards (called community cards) on the table in three steps: 3 cards (flop), 1 card (turn) and 1 card (river)), find:

a) how many ways pocket cards can be dealt to one player;
b) how many ways pocket cards can be dealt to two players;
c) how many ways pocket cards can be dealt to n players;

Solution:

a) Obviously, the number of all combinations is

$$C_{52}^2 = \frac{51 \cdot 52}{2} = 1326.$$

b) For two players, we have to count the double 2-size combinations from 52 cards. These have a ((xy)(zt)) form, with x, y, z and t being mutually distinct. Note that in a simple combination the order does not count, while in a double combination the order

213

does count (the first element (xy) stands for the cards dealt to the first player and the second element (zt) stands for the cards dealt to the second player).

In other words, the double combination ((ab)(cd)) is different from ((cd)(ab)).

To perform a correct count, we hypothetically unfold these combinations by fixing a combination for the first player and unfolding all possible combinations for the second from the cards left, which number C_{50}^2.

For the first player we have C_{52}^2 possible combinations; therefore, the number of possible double combinations is $C_{52}^2 \cdot C_{50}^2$ (for an added exercise, do the numeric calculation).

c) For n players, the count also consists of similar successive multiplications. Let us observe some partial results:

– For two players, we have found $C_{52}^2 \cdot C_{50}^2$.

– For three players, we have $C_{52}^2 \cdot C_{50}^2 \cdot C_{48}^2$.

– For four players, we have $C_{52}^2 \cdot C_{50}^2 \cdot C_{48}^2 \cdot C_{46}^2$.

Note the last factor of each partial result and observe that:

$50 = 52 - 2 \cdot 1$ (for two players)

$48 = 52 - 2 \cdot 2$ (for three players)

$46 = 52 - 2 \cdot 3$ (for four players).

By generalizing, we find the formula of all possible combinations of distribution to n players is:

$$C_{52}^2 \cdot C_{50}^2 \cdot \ldots \cdot C_{52-2(n-1)}^2 \text{ or, written differently, } \prod_{i=1}^{n} C_{52-2(i-1)}^2 .$$

15) In Texas Hold'em, how many ways can the five community cards be distributed?

Solution:

If we look at this problem from a neutral observer's perspective (someone who does not participate at the game), the combinations are generated from all 52 cards, so their number is

$$C_{52}^5 = \frac{48 \cdot 49 \cdot 50 \cdot 51 \cdot 52}{2 \cdot 3 \cdot 4 \cdot 5} = 4 \cdot 49 \cdot 5 \cdot 51 \cdot 52 = 2598960 .$$

If we look at this problem from a participating player's perspective (a player who has already seen his or her pocket cards), then the combinations are generated from 50 cards and their number is

$$C_{50}^5 = \frac{46 \cdot 47 \cdot 48 \cdot 49 \cdot 50}{2 \cdot 3 \cdot 4 \cdot 5} = 46 \cdot 47 \cdot 2 \cdot 49 \cdot 10 = 2118760.$$

16) In Texas Hold'em, how many ways can the flop cards be distributed?

Solution:
From a neutral observer's perspective, the three flop cards can be distributed in $C_{52}^3 = \dfrac{50 \cdot 51 \cdot 52}{2 \cdot 3} = 50 \cdot 17 \cdot 26 = 22100$ ways.

From a participating player's perspective (a player who already has his or her two cards), the number of combinations for the flop is

$$C_{50}^3 = \frac{48 \cdot 49 \cdot 50}{2 \cdot 3} = 8 \cdot 49 \cdot 50 = 19600.$$

17) In Texas Hold'em, how many ways can the pocket cards be dealt to a player so that they contain at least one ace?

Solution:
Denoting an ace by A (A can be A♣, A♠, A♦, A♥), we can count the combinations of (Ax) form, with $x \neq A$ and (AA) form, the results following to be added.

For (Ax): A takes four values and x takes $52 - 4 = 48$ values, so the number of combinations is $4 \cdot 48 = 192$.

For (AA): The number of combinations is $C_4^2 = 6$.

Then, the number of combinations containing at least one ace is $192 + 6 = 198$.

18) In Texas Hold'em, how many ways can the pocket cards be dealt to a player so that they to contain two identical symbols (two cards of the same suit)?

Solution:

For a specific symbol S, the combinations to be counted have an

(SS) form and their number is $C_{13}^2 = \dfrac{12 \cdot 13}{2} = 78$.

We have four symbols, so the total number of combinations is $4 \cdot 78 = 312$.

19) You are participating in a Texas Hold'em game and are dealt the 7♣ and the 8♦. How many ways can the flop cards be distributed so that they contain: a) exactly two clubs; b) a minimum of two clubs?

Solution:

a) A flop combination containing exactly two clubs has a (CCx) form, with $x \neq C$ (C = clubs); C takes $13 - 1 = 12$ values (a clubs card from a total of 13 is in the hand); the set of values x can take has $52 - 2 - 2 - 10 = 38$ elements (we have subtracted the two pocket cards, the two CC cards from the flop combination and the ten clubs left).

(CCx)=(CC)(x)

The number of combinations is $C_{12}^2 \cdot 38 = \dfrac{11 \cdot 12}{2} \cdot 38 = 2508$.

b) Let us calculate the number of combinations containing three clubs. These have a (CCC) form and number

$C_{12}^3 = \dfrac{10 \cdot 11 \cdot 12}{2 \cdot 3} = 220$.

In total, we have $220 + 2508 = 2728$ combinations containing at least two clubs.

20) In Texas Hold'em, how many possible hands containing a pair (any) can a player be dealt?

Solution:

For a specific pair (PP) (*PP* are two different cards of same value), there are $C_4^2 = 6$ possible combinations (*P* takes four values).

We have thirteen values for all cards; therefore, the total number of combinations is $13 \cdot C_4^2 = 13 \cdot 6 = 78$.

21) At a slot machine (a machine with 3, 4 or 5 reels, each reel have the same number of different symbols, with 1, 2 or 3 winning lines; after an aleatory spin of the reels, a combination of symbols that may win or not occurs on the winning line) with three reels, seven symbols and one winning line:

a) How many possible combinations of symbols can occur on the winning line?

b) How many of these combinations contain three identical symbols?

c) How many of these combinations contain exactly two identical symbols?

Solution:

a) Because there are seven symbols on each of the three reels, we have $7 \cdot 7 \cdot 7 = 343$ possible combinations.

b) For a specific symbol S, we have a single combination containing 3 S-symbols, namely (SSS).

In total, we have $7 \cdot 1 = 7$ combinations.

c) For a specific symbol S, the combinations to be counted have the form:

(SSx), $x \neq S$, in number $1 \cdot 1 \cdot 6 = 6$

(SxS), $x \neq S$, in number $1 \cdot 1 \cdot 6 = 6$

(xSS), $x \neq S$, in number $1 \cdot 1 \cdot 6 = 6$.

Obviously, the order of elements has been taken into account.

We have no common combinations among these combinations (such a combination would have an (SSS) form, but $x \neq S$), so we are allowed to add the three numbers and we find 18.

We have seven symbols; therefore, the total number of searched combinations is $7 \cdot 18 = 126$.

22) How many possible ways can three dice fall after they have been rolled (with respect to the numbers shown on their superior side)?

Solution:

The problem is similar to point a) of the previous problem (we use a die in place of a reel and the die numbers in place of symbols on the reel).

Each die has six numbers, so the three dice can show $6 \cdot 6 \cdot 6 = 216$ possible combinations after rolling (one certain number on each die—the order does count).

23) How many possible ways can three dice fall after they have been rolled so that the sum of the numbers shown is 15?

Solution:

The problem reverts to partitioning of the number 15 in 3 terms, such that each term is less than or equal to 6, by taking into account the order of terms.

So as not to miss any partition and to avoid duplication, we will progressively increase the terms from left to right, starting with 1 on the first place:

$1 + 1 + 13$ – inconvenient $(13 > 6)$
$1 + 2 + 12$ – inconvenient $(12 > 6)$

No combination containing 1 is convenient because the remaining fourteen cannot be partitioned into two numbers that are less than or equal to 6:

$2 + 2 + 11$ – inconvenient
$2 + 3 + 10$ – inconvenient
...
$2 + 6 + 7$ – inconvenient
Continuing with 3 on the first place:
$3 + 2 + 10$ – inconvenient
...
$3 + 5 + 7$ – inconvenient
$3 + 6 + 6$ – convenient
Continuing with 4 on the first place:
$4 + 2 + 9$ – inconvenient
...
$4 + 4 + 7$ – inconvenient
$4 + 5 + 6$ – convenient
$4 + 6 + 5$ – convenient
Continuing with 5 on the first place:
$5 + 2 + 8$ – inconvenient
$5 + 3 + 7$ – inconvenient
$5 + 4 + 6$ – convenient
$5 + 5 + 5$ – convenient

5 + 6 + 4 – convenient
Finally, 6 on the first place:
6 + 2 + 7 – inconvenient
6 + 3 + 6 – convenient
6 + 4 + 5 – convenient
6 + 5 + 4 – convenient
6 + 6 + 3 – convenient.
In total, we have ten different combinations whose elements make a total of 15, namely, 3-6-6, 4-5-6, 4-6-5, 5-4-6, 5-5-5, 5-6-4, 6-3-6, 6-4-5, 6-5-4, 6-6-3.

As we also mentioned in other application, the term *combination* is abusive in this case because the order does count.

24) In a race among eight competitors, how many ways can the first three places be taken? How many ways can all eight competitors be ranked?

Solution:

We deal here with 3-size arrangements from an 8-size set, so the number of possible ways of taking the first three places is $A_8^3 = 6 \cdot 7 \cdot 8 = 336$ (Observation: This number also stands for the last three places or any other three places)

The all eight competitors can be arranged (permutated) in $A_8^8 = 8! = 5040$ ways.

25) How many ways can a sport betting ticket be filled, if on the ticket there are four matches and the results to bet on are 1, X or 2 (1 – hosts win, X – draw, 2 – visitors win)?

Solution:

The problem is similar to the slots one or dice one.

Four matches of three results each generate $3^4 = 3 \cdot 3 \cdot 3 \cdot 3 = 81$ possible (ordered) combinations.

Unsolved applications

1) Calculate: C(8, 3), C(11, 8), C(17, 5), C(19, 12), C(25, 7), C(41, 16), C(52, 43), A(7, 2), A(11, 7), A(15, 12), A(23, 4), A(31, 15), A(40, 11), A(47, 3).

2) In a 6/49 lottery, find:
a) the number of drawn variants possible containing numbers 5 and 17;
b) the number of drawn variants possible containing numbers 2, 7 and 10;
c) the number of drawn variants possible containing exactly three even numbers;
d) the number of drawn variants possible containing a minimum of three even numbers;
e) the number of drawn variants possible containing only number that are larger than 12;
f) the number of drawn variants possible containing at least three numbers larger than 15.

3) In a 5/40 lottery, find:
a) the number of possible played variants containing a minimum of three even numbers;
b) the number of drawn variants possible containing a minimum of three even numbers;
c) the number of possible variants played containing exactly two numbers less than 15;
d) the number of drawn variants possible containing exactly two numbers less than 15;
e) the number of possible played variants containing a minimum of two numbers less than 17;
f) the number of drawn variants possible containing a minimum of two numbers less than 19.

3) In a 52-card classical poker game, find:

a) the number of possible hands a player can be dealt that contain at least one K (king);

b) the number of possible hands a player can be dealt that contain exactly three ♦ symbols (diamonds);

c) the number of possible hands a player can be dealt that contain exactly three identical symbols (any);

d) the number of possible hands a player can be dealt that contain two pairs (any; no full house, no quads, but exactly two different pairs);

e) the number of possible hands a player can be dealt, only containing cards with a value larger than 9;

f) the number of possible hands a player can be dealt that contain four cards of same value (quads).

5) Same problem for a 32-card classical poker game.

6) In Texas Hold'em Poker, find the number of possible hands a player can be dealt , containing:

a) one 2 and one 5;

b) at least one 2 or at least one 5;

c) one less than 10 card and one larger than 10;

d) cards having different symbols (unsuited).

7) In Texas Hold'em, from the perspective of a neutral observer, calculate how many ways the five community cards can be distributed such that:

a) to contain exactly three diamonds (♦);

b) to contain a minimum of three diamonds;

c) to contain exactly three identical symbols (any);

d) to contain a minimum of three identical symbols;

e) to contain exactly three cards with same value (triple, or three of a kind);

f) to contain a minimum of three cards of same value (3 or 4);

g) to be all consecutive (for example: A 2 3 4 5 or 3 4 5 6 7);

h) to contain exactly four consecutive cards.

8) In Texas Hold'em, you are participating as a player and you were dealt 2♦ and J♠. How many ways can the flop cards be distributed so that they contain:

    a) exactly two clubs;

    b) a minimum of two clubs;

    c) 2's or J's (however many of them);

    d) 2's and J's (however many);

    e) a pair of 2's (no triple);

    f) a pair of J's

    g) a pair of 2's or a pair of J's (no triple).

9) At a slot machine with five reels, eight symbols and one winning line:

    a) How many combinations can occur on the winning line?

    b) How many of them contain exactly three identical symbols (any)?

    c) How many of them contain a minimum of three identical symbols (any)?

10) How many ways can four dice fall after being rolled? How many ways can four dice fall after being rolled, such that the sum of numbers shown is 17?

11) How many ways can a sport betting ticket be filled, if on the ticket are seven matches and the result of the bet is the total number of scored goals: 0 – 3, 4 – 6, 7 – 10 and >10 (four variants)?

BEGINNER'S CALCULUS GUIDE

Introduction

The previous chapters presented the probability notion in all its interpretations in lay language as well as at the mathematical and philosophical level.

As can be seen in these sections, no interpretation variant can omit the mathematical model given by probability theory.

When we speak about probability calculus, we refer strictly to the mathematical calculus tools provided by this theory, and exclude other subjective interpretations.

Although this calculus often assumes hypothetical approximations or pure calculation, in the end or even from the start, the numerical results thus obtained are much more relevant than any other subjective estimation based on intuition or on nonmathematical interpretations of probability.

This chapter is a guide to the calculus of numerical probabilities, and is structured so that it can be used by persons with a minimal mathematical background.

Although theoretically the guide can be studied and used without running through the mathematical chapter as a preliminary— because the formulas used in the applications are presented again for review before being applied and the presentation of solutions is algorithmic— we consider that a minimal knowledge of the classical definition of probability, probability field, operations with events and relations between events to be necessary.

Also, the combinatorial calculus represents a main tool that is used consistently in the applications presented, and understanding and applying this calculus requires reading the chapter titled *Combinatorics* before proceeding with these applications.

Besides this basic knowledge of probability theory and combinatorics, the only requirement for the reader is to have a good command of the four arithmetic operations between real numbers and of basic algebraic calculus.

These limitations in knowledge are possible because this guide deals only with finite or at the most discrete cases as the basis for these applications.

Most of the applications presented here come from games of chance, where we deal only with finite probability fields.

Theoretically, any probability calculus problem, no matter how complex, can be unfolded in successive elementary applications that use basic formulas, but most of the time finishing the calculus can be very laborious or even impossible, not to mention the high risk of the occurrence of errors during a long succession of calculations.

The use of combinatorics and even of classical probability repartitions can often solve such problems simply and elegantly, whereas the step-by-step approach is much too laborious and is predisposed to calculation errors.

If we composed a list of the minimal knowledge required by the reader who wants to solve finite probability applications by studying this guide, it would look as follows:

Previous background (from school):
– Operations with real numbers: addition, subtraction, multiplication, division, powers, order of operations, operations with fractions, reductions;
– Algebraic calculus: expanding the brackets, multiplication of expressions within brackets, raising an expression to a power, formulas of shortened calculus, factoring out, reduction of algebraic fractions;

Combinatorics knowledge (from school or the chapter titled *Combinatorics)*:
– Definition of permutations, arrangements and combinations;
– General formulas of permutations, arrangements and combinations;
– Combinatorial calculus procedures: properties (formulas), partitioning;
– Models of solved applications;
– Solving as many as applications as possible.

Basic probability knowledge (from school or the chapter titled *Probability Theory Basics*):

– Operations with sets: intersection, union, difference, complementary;

– Events, operations with events as sets;

– Elementary events, incompatible events, independent events, mutually exclusive events;

– Classical definition of probability;

– Probability field.

Of course, the list can be extended depending on each reader's option; thus, we recommend that the wider the image of theoretical notions and results are, the surer and much better guarded against theoretical or calculation errors their application will be.

Inversely, the list can be reduced (but not by much) because the solution of many applications (especially those from the beginning) contain additional explanations about the notions involved in framing a problem and in the mathematical models used.

Besides this list of required mathematical knowledge, an important component of the ensemble of necessary skills is the ability to observe.

The correct framing of a problem, establishing the probability field to operate within and the relations between various events are a matter of solver's ability to observe.

This initial stage of the solving algorithm is essential in solving an application and finding the final numerical result.

But the ability to observe is not a native skill and does not result from a previous mathematical education. It can be acquired at any time through unceasing exercise.

This is also the reason for including in this guide a collection of solved applications with detailed explanations and instructions.

The complexity and difficulty of the applications grows progressively and their solutions follow exactly the general algorithm for solving presented in the next section.

The didactic goal of this guide is to enable the reader to solve any finite probability application alone or by consulting the guide.

At the end of this introduction, to achieve the best didactic results, we recommend that the reader do the following things:

– Read the sections that explain the probability notion in the chapters titled *What Is Probability?* (the sections on *Probability—the word, Probability as a limit, The probability concept*).

– Read at least once all definitions in the list of requirements of mathematical knowledge outlined earlier.

– Follow exactly the general algorithm for solving for any application and do not skip any stage, even if a particular application appears easy (do not pass directly to numerical calculus without framing the problem and establishing the probability field).

– Do revise the long or complex combinatorial calculations; this is where errors often occur.

– Do not try to apply a previous solving scheme to a similar application at any cost; the probability fields may be different, as can the events to be measured or the questions to be answered, even if their descriptions are similar or even identical.

– Try to memorize over time the formulas used to solve the applications.

– Revise the calculations and the entire solution algorithm each time the final numerical result seems too low or too high, but do not turn this into a general criterion for establishing the existence of an error (intuition can play bad tricks in probability).

– Do not approach the unsolved applications (except those recommended at the end of various sections) until you have run through the entire calculus guide and its solved applications.

– Do not stop your study if from the beginning you feel that solving the applications is beyond your comprehension; just take a break and resume reading later, when your concentration is higher. Read a paragraph several times if needed and come back to the information from previous chapters any time you consider it necessary.

The general algorithm of solving

Every solution of a probability application submits to a basic algorithm, which basically ensures the correctness of framing and approach to the calculus problem and of the application of the theoretical results as well.

Even though the methods of solving a problem can be multiple, all procedures are applied on the basis of this general algorithm, which is valid for any finite or discrete probability application.

The solution algorithm consists of three main stages:

1) *Framing the problem*
– Establishing the probability field attached to an experiment;
– Textually defining the events to be measured;
– Establishing the elementary events that are equally possible;
– Observing the independent, nonindependent and incompatible events;
– Necessary idealizations.

2) *Establishing the theoretical procedure*
– Choosing the solving method (step by step or condensed);
– Selecting the formulas to use.

3) *The calculus*
– Numerical calculus;
– Combinatorial calculus;
– Eventual approximations;
– Probability calculus (applying the formulas).

Framing the problem

This first stage of the solution algorithm is very important.

Although it does not include the probability calculus that is required by each application, it establishes the framework for this calculus by showing the optimal mathematical model that makes the correct application of theoretical results possible and ensures that relevant numerical results are acquired in the end.

We saw that the probability of an event makes sense from a mathematical point of view only if that event belongs to a Boolean structure, namely, a field of events.

If we consider the set Ω of all possible outcomes or the sample space of an experiment, then the set of events associated with that experiment is included in or equal to $\mathcal{P}(\Omega)$ and is a Boole algebra.

If we specified the set Ω and this set is finite or discrete, we have also specified the associated field of events.

Examples:

1) In the experiment involving tossing a coin, the sample space is $\Omega = \{H, T\}$ (H – heads, T – tails), and the field of events is:
$\mathcal{P}(\Omega) = \{\phi, \{H\}, \{T\}, \Omega\}$.

2) In the experiment involving tossing two coins, the sample space may be $\Omega = \{(H,T), (T,H), (H,H), (T,T)\}$ if we take into account the outcomes for each coin (we deal in this case with a set of ordered pairs), or may be $\Omega' = \{(HT), (HH), (TT)\}$, if we take into account the cumulative outcomes for both coins (it is a set of unordered pairs, namely combinations, in which order does not count).

Although the set Ω' stands for a set of outcomes that covers all possibilities, choosing the field of events $\mathcal{P}(\Omega')$ as basis for framing the application is not correct.

As we will see further, the elementary events of this field cannot be considered equally possible.

228

In applications, specifying the sample space by enumerating its elements is not absolutely necessary because we are most interested in the number of these elements rather than in being able to see them.

3) Three persons are randomly chosen from a group of 100 with certain specified characteristics. The probability for at least one chosen person to have certain characteristics is required.

In such a problem, denoting all possible outcomes of the experiment and unfolding their set is useless. This set has only to be established and imagined.

The number of its elements is given by all 3-size combinations from 100 elements, namely, $C_{100}^3 = 161700$.

Often, the field of events is not that easy to view, as can be seen in the next example. In these cases, the field of events must be rebuilt so that the event whose probability we are looking can belong to it.

4) A bus comes regularly at a bus station every twenty minutes, starting at 8 a.m. and stopping for 1 minute. A person coming to the bus station from a long distance may get there any time between 8 a.m and 9 a.m. What is the probability of that person finding the bus during the minute it is stopped at the station?

Unlike the gambling applications, where the entire sample space can be automatically visualized, in this experiment we must rebuild the events that make up this set by starting from the event to be measured.

Because the bus stops for 1 minute, we split the 1-hour interval (during which the person can arrive at the bus station) into 60 minutes. Thus, the minute stands for the unit to be taken into account when establishing the elementary events:

e_i – the person arrives in the bus station in the minute i

$(i = 1, ..., 60)$ represent the elements of set Ω.

```
   1m      20 m     1m      20 m     1m      17 m
  |...|..................|...|..................|...|...............|
8 a.m.                                                      9 a.m.
```

229

According to the diagram above (which divides the 8 a.m. – 9 a.m. time interval into minutes, with the 1-minute intervals representing the stopping times), the event to be measured is $e_1 \cup e_{22} \cup e_{43}$, which consists of three equally possible elementary events (respectively, *the person arrives in the minute 1, the person arrives in the minute 22*, or *the person arrives in the minute 43*).

5) Some situations are so vaguely phrased that the problem cannot be solved by applying probability theory because the field of events cannot be configured or even imagined.

Questions like "What is the probability for the Sun to not rise tomorrow?" or "What is the probability for an extraterrestrial base to lie on the bottom of the ocean?" cannot be framed with respect to a field of events because the information that can be taken into account cannot be quantified or is missing entirely.

In these cases, the experimental context is missing, as is the Boolean structure that can include them.

Estimating the probability of events isolated from any mathematical context cannot rely on the tools of theory and thus cannot be a matter of the practical application of this guide.

Any quantitative or qualitative reference to the degree of belief in the occurrence of such events is purely subjective and has nothing to do with mathematical probability.

Mathematical probability is uniquely determined by the probability field associated with an experiment, namely, the triple (Ω, Σ, P), where:

Ω is the set of all possible outcomes;

Σ is the field of events (included in $\mathcal{P}(\Omega)$);

P is the probability function defined on Σ.

We already saw in the section titled *Relativity of probability* how we can get different numerical values of the function *P* for the same event if it is framed within different probability fields.

Establishing the probability field means specifying each of its three components.

In finite or discrete applications, this reverts to specifying the set of equally possible elementary events, because:

230

– $\Sigma = \mathcal{P}(\Omega)$, so the field of events is generated by the sample space;

– The theorem of uniqueness of measures ensures the uniqueness of the probability function on a field of events generated by a set of equally possible elementary events.

Thus, in all practical applications of this guide, establishing the probability field means establishing the set Ω of all possible outcomes, or the set of elementary events.

Most of time we are interested in the number of elements of this set and not in its visualization.

Choosing the field of events when it is not immediately visible is influenced by the textual definition of the event to be measured.

Therefore, the description of the event through words becomes very important to framing the event within the appropriate probability field.

Any vague definition or definition containing inconsistent information can result in a simplistic or irrelevant framing.

The use of as many words as possible to describe an event to be measured is recommended, even if it is excessively lengthy or even modifies the event's description from the initial statement of the problem.

For example, at a card game, to find the probability of an opponent holding an ace, the following phrasings of the event's definition are not at all indifferent for calculus:

– the opponent holds an ace
– the opponent holds an ace after I saw my own cards, or
– the opponent holds an ace after I saw my own cards and another three cards from the deck.

Each event described in these phrases belongs to a different field of events, even if the partial description *the opponent holds an ace* represents the simplest way to state the final event.

In fact, it is about the initial information taken into account, which transposes itself into choosing a certain field of events and, implicitly, a certain probability field.

Among all descriptions of an event, we must choose the one that best corresponds to the problem and that contains the most relevant information.

We have already outlined the first two actions of solving a finite or discrete application: establishing the set of elementary events and textually defining the events to be measured.

It is ideal for the elementary events to be equally possible (in the generally accepted understanding of this term: none of event's occurrences is favored by the conditions in which the experiment is performed.).

Giving the quality of *equally possible* assumes a necessary approximation that makes the probability calculus applicable.

In gambling applications, this idealization is ensured and unanimously accepted, and is justified by the aleatory technical conditions in which the game runs.

In other applications, this type of idealization is made as an additional assumption of the hypothesis.

In the previous example 4 (the probability of catching a bus at the station), we considered the events *the person arrives in the bus station in the minute i* as being equally possible elementary events.

In this way, we make the assumption that there are no minutes among the 60 in which the person can arrive having a higher measure that in others; that is, there are no favored minutes.

Thus, we have neglected the fact that the departure hour, the transportation means used and other factors can influence the arrival minute, which is an approximation.

Concerning the possibility of choosing the elementary events, it is not unique. We can just as well divide the one hour interval into seconds instead of minutes, and consider the events *the person arrives in the bus station in the second* i (with i = 1 ,..., 3600) as elementary.

Such a choice is not incorrect and leads to the same final numerical result for the probability of the event to be measured.

We can choose any set of elementary events as long they are equally possible.

In the various less or more complex applications, the necessity of other idealizations besides *equally possible* may appear for assumptions or approximations that are necessary to frame the problem and application of calculus formulas to be allowed.

Depending on requests and hypotheses, in some problems we can consider a complete system of events (a finite or countable family of mutually exclusive events whose union covers the set of possible outcomes) instead of the sample space.

After we build a correct image of the probability field to operate within (we established the elementary events, the sample space or a complete system of events), it is also useful to observe the relations between the various events involved in a problem's solution:

- Inclusion relations ($A \subset B$);
- Incompatible events ($A \cap B = \phi$);
- Independent events (events whose occurrence is not mutually influenced, according to definition, $P(A \cap B) = P(A) \cdot P(B)$);
- Nonindependent events (events that are not independent, that is, their occurrence is influenced at least in one direction).

It is important to observe how the events to be measured can be decomposed (as $A = B \cup C$), in order to apply the properties of probability and the calculus formulas.

Some applications may contain events that are too complex to be decomposed into elementary events, and in these cases the application of the basic formulas and the classical definition of probability are no longer practical if they are based only on observations.

In such applications, framing the problem consists of finding the type of classical probability distributions that matches the hypothesis.

We now present a few simple exercises with the goal of making the correct framing of a problem by establishing the probability field, the events involved and the relations between them.

Exercises and problems

1) At a European roulette (with thirty-seven numbers from 0 to 36), find the probability of occurrence of an uneven number after a spin.

Answer:
The sample space is $\{0, 1, 2, ..., 36\}$.
The events $e_i = \{$occurrence of number i$\}$, $i = 0, ..., 36$, are the elementary events of the experiment of spinning the roulette wheel.
They are equally possible (this is a necessary idealization). Each event has the probability 1/37.
The event to be measured is A – *occurrence of an uneven number*. This is a compound event, which can be decomposed as $A = e_1 \cup e_3 \cup ... \cup e_{35}$ (in eighteen elementary events).
The elementary events are mutually exclusive; therefore, we have $P(A) = P(e_1) + P(e_3) + ... + P(e_{35}) = 18 \cdot 1/37 = 18/37 = 0.48648$.
In other words, from the total of thirty-seven equally possible outcomes, eighteen are favorable for the event *A* to occur, implying an 18/37 probability according to the classical definition of probability.

2) Two dice are rolled simultaneously. Calculate the probability for the sum of the points shown on both dice to be greater than 7.

Answer:
An elementary event is represented by a 2-size combination of numbers of the two dice.
The set of elementary events is then
$\{(a, b) | a \in \{1, 2, 3, 4, 5, 6\}, b \in \{1, 2, 3, 4, 5, 6\}\}$, which is a set of combinations with 6 x 6 = 36 elements.
a stands for the number shown on the first die and *b* for the number shown on the second.
Any such combination (a, b) is possible in the same measure.
The event to be measured is A: $a + b > 7$.
All the variants that are favorable for this inequality are:
$2 + 6, 3 + 5, 3 + 6, 4 + 4, 4 + 5, 4 + 6, 5 + 3, 5 + 4, 5 + 6, 6 + 2,$
$6 + 3, 6 + 4, 6 + 5, 6 + 6$, for a total of 14.

Observe that the order has been taken into account (both combinations $a + b$ and $b + a$ have been counted as different).

Each favorable combination is an elementary event and their union is event A.

We then have $P(A) = 14 \times 1/36 = 14/36 = 7/18 = 0.38888$.

An incorrect framing of this problem is one in which we would consider the set of elementary events as the set of possible sums of the points shown on dice: {sum 2, sum 3, sum 4, sum 5, sum 6, sum 7, sum 8, sum 9, sum 10, sum 11, sum 12}, with eleven elements.

Event A is the union of five elementary events (sum 8, sum 9, sum 10, sum 11 and sum 12).

Although it seems to be an easy choice and does not take into account the order (but the cumulative results of the two dice), it is not correct because the respective events cannot be considered equally possible.

For example, the event *sum 2* can occur in only one way ($1 + 1$, namely number 1 on both dice), while the event *sum 5* can occur in four ways ($2 + 3, 3 + 2, 1 + 4, 4 + 1$).

This makes impossible an *equally possible* type of idealization for these events.

In such a field of events, calculating the probability of event A as $5/11$ is incorrect because the classical definition of probability is valid only for equally possible events. Keep this error example in mind.

3) In a Texas Hold'em Poker game (a 52-card deck is used and each player is dealt two cards), calculate the probability for a certain player to be dealt a pair (two cards of same value).

Answer:

All 52 cards are unknown (there are no viewed cards at the moment of calculation). The elementary events are the occurrences of various 2-size combinations of cards taken from the 52.

We can consider without reservation the approximation that these events are equally possible (no combination is favored).

The total number of these combinations is $C_{52}^2 = 1326$. This is the number of elements of the sample space.

Denote by A the event *player is dealt a pair*.

If we denote by:

A_2 : player is dealt a pair of 2's

A_3 : player is dealt a pair of 3's

A_4 : player is dealt a pair of 4's

A_5 : player is dealt a pair of 5's

A_6 : player is dealt a pair of 6's

A_7 : player is dealt a pair of 7's

A_8 : player is dealt a pair of 8's

A_9 : player is dealt a pair of 9's

A_{10} : player is dealt a pair of 10's

A_J : player is dealt a pair of J's

A_Q : player is dealt a pair of Q's

A_K : player is dealt a pair of K's

A_A : player is dealt a pair of A's,

the above events (totaling thirteen) are mutually exclusive and we have: $A = A_2 \cup A_3 \cup ... \cup A_{10} \cup A_J \cup A_Q \cup A_K \cup A_A$.

Probability P(A) is then the sum of the probabilities of these events.

To calculate the probability of A_2 , we must count the combinations of (2, 2) type (the first card is a 2, the second is also a 2).

Because we have four cards with the value 2 (namely 2♠, 2♣, 2♥, 2♦), the number of these combinations is $C_4^2 = 6$.

Probability of A_2 is then 6/1326 (the number of favorable combinations divided to the number of all possible combinations).

Similarly, we can calculate the probabilities of the remaining events, which are the same.

We then have P(A) = 13 x 6/1326 = 0.05882.

4) You are participating in a Texas Hold'em Poker game with three opponents. You are dealt a 2 and a 3 and the flop cards are A, 5, 7.

Calculate the probability for at least one opponent to hold an ace.

Answer:

By focusing on one opponent, the first problem to solve is finding the probability for that opponent to hold at least one ace.

Here, too, the equally possible elementary events are the occurrences of the various 2-size combinations of cards in the opponent's hand.

But these combinations are no longer generated from the 52 cards, but from 47 cards because you have seen five of them (your own two cards plus the three flop cards), so they cannot occur elsewhere.

So, the probability field here is different than the field in previous exercise.

Obviously, we are not interested in the symbols of these cards, but in their values.

The set of all possible outcomes has then $C_{47}^2 = 1081$ elements.

The event to be measured (the opponent holds at least one ace) can be written as A – *the opponent holds a combination of (Ax) type, where x may be any card except the ones viewed.*

The next step is to calculate the number of these combinations and to divide this number by 1081 in order to get P(A).

The initial problem asks for the probability of event B – *at least one opponent holds an ace.*

Event *B* is the union of the following mutually exclusive events:

B_1 – a single opponent holds at least one ace;

B_2 – exactly two opponents hold at least one ace each;

B_3 – all three opponents hold at least one ace each;

After calculating the probabilities of these three events, they must be added to get P(B).

Complete solutions of such applications are presented in the following sections.

5) In a classical poker game (each player is dealt five cards) with a 32-card deck (from 7 upward), find the probability for a player to be dealt a straight first (the five dealt cards are consecutive).

Answer:

The elementary events we consider here are the occurrences of the various 5-size combinations of cards from the 32 in a player's hand. Their number is $C_{32}^5 = 201376$.

The event to be measured is A – *occurrence of a straight*.
If we denote by:

A_1 – player is dealt a combination of (7 8 9 10 J) type

A_2 – player is dealt a combination of (8 9 10 J Q) type

A_3 – player is dealt a combination of (9 10 J Q K) type

A_4 – player is dealt a combination of (10 J Q K A) type,

then *A* is the union of these mutually exclusive events.

Each of the events A_i is a composition (in the sense of union) of elementary events corresponding to the occurrences of the respective combinations.

The number of these elementary events is given by the number of combinations of that type, namely, 4 x 4 x 4 x 4 x 4 = 1024 for each event A_i.

We then have P(A) = (4 x 1024)/201376 = 0.02034.

6) Ten devices of same type are put in operation: three come from factory F_1, five from factory F_2 and two from factory F_3. The devices are the subject of a test. The three from the first factory pass the test with probability 0.9, the five from the second factory with probability 0.75 and the two from the third factory with probability 0.85. A device is randomly chosen. What is the probability for the chosen device to pass the test?

Answer:

In this type of problem, establishing the elementary events is not required.

If we denote by A_i – *the device comes from factory* F_i, then $\{A_i\}_{i=1,2,3}$ is a complete system of events (it covers through a disjointed union the entire sample space of the experiment of choosing the device).

We have $P(A_1) = 3/10$, $P(A_2) = 1/2$ and $P(A_3) = 1/5$.

Denote by A the event to be measured *the device passes the test* and by $A|A_i$ the events *the device passes the test if it comes from factory* F_i, $i = 1, 2, 3$.

The events A and A_i are nonindependent for any i.

We have, according to the hypothesis, the following conditional probabilities:

$P(A|A_1) = 0.9$, $P(A|A_2) = 0.75$, $P(A|A_3) = 0.85$.

We apply the formula of total probability and we find

$$P(A) = \sum_{i=1}^{3} P(A_i)P(A|A_i) = 163/200 = 0.815.$$

7) Three persons are independently flipping one coin each. Find the probability of all three coins showing heads.

Answer:

Denoting by A – *the first coin shows heads*, B – *the second coin shows heads* and by C – *the third coin shows heads*, observe that A, B and C are mutually independent events.

The event to be measured is $A \cap B \cap C$.

We then have:

$P(A \cap B \cap C) = P(A) \cdot P(B) \cdot P(C) = 1/2 \cdot 1/2 \cdot 1/2 = 1/8$.

Such an approach considers three independent experiments, each having its own sample space $\{H, T\}$ (heads and tails).

The problem can be also framed by using a single probability field, if we consider flipping the three coins as a single experiment in which the set of elementary events is identified with the set of combinations (x, y, z), with $x, y, z \in \{H, T\}$.

The number of its elements is $2^3 = 8$.

We have a single favorable combination corresponding to events *all three coins show heads*, namely (H, H, H), so its probability is 1/8.

8) A pair of dice is rolled six times. What is the probability of getting a total of 7 points exactly three times?

Answer:

This type of application assumes identifying a classical probability scheme to match it and its application.

In this case, it is about a Bernoulli scheme, which must be applied for n = 6 and m = 3.

For an individual experiment of rolling two dice, event A – *we obtain a total of 7 points* has the probability 6/36 (six favorable combinations, corresponding to sums 1 + 6, 2 + 5, 3 + 4, 4 + 3, 5 + 2, 6 + 1, from 6 x 6 = 36 possible).

We then have p = 1/6.

According to the Bernoulli scheme, the searched probability is

$$P_{3,6} = C_6^3 \left(\frac{1}{6}\right)^3 \left(\frac{5}{6}\right)^3.$$

These solved applications only hint at framing the problem and not at the probability calculus itself.

Applications containing complete solutions and detailed calculations of probabilities are presented in the following sections.

We propose as an exercise framing the following problems (without a complete solution):

1) A person randomly chooses a card from a 52-card deck, without seeing any card. What is the probability for that card to be 7♥?

2) A person randomly chooses a card from a shuffled 52-card deck. During shuffling, a card is dropped face up and seen to be the J♣.

What is the probability of the chosen card being 8♦?
What is the probability for the chosen card to be an *A*? How about a *J* ?

3) Four dice are rolled. Find the probability of the dice showing four identical numbers.

4) You are participating in a Texas Hold'em Poker game and you are dealt 5♠, Q♥. Find the probability for the flop cards to contain a queen. Find the probability for a specific opponent to hold a queen.

5) We have two urns. One contains three white balls and four black balls and the other contains four white balls and five black balls. An urn is randomly chosen and a ball is drawn from it. What is the probability the drawn ball will be white?

6) A coin is flipped seven times. Find the probability for heads to come up at least once.

7) In a 6/49 lottery system, find the probability for the drawn variant to contain five previously fixed numbers.

Establishing the theoretical procedure

To establish the theoretical procedure through which an application is solved means choosing a solving method and selecting the mathematical formulas to be used in the probability calculus itself.

This selection is always made after the problem is properly framed. No formula can be applied without having first defined the ensemble of conditions (hypotheses) that match the mathematical model that generated the respective formula.

This is why respecting the chronological order of the two stages (framing the problem and establishing the theoretical procedure) is not just a recommendation, but a logical necessity.

Methods of solving

A probability calculus problem may have several solving methods (ways) that lead to the same correct result. This happens because probability theory is a consistent and rigorous theory from a mathematical point of view.

The numerous possible solving methods that use the basic theoretical results can be grouped into two main categories: the step-by-step methods and the condensed methods.

The solutions of finite type applications all use these solving methods, either individually or combined.

Obviously, depending on the application, other specific methods might appear that use more complex theoretical results, and these cannot be framed in the two main categories.

A solution might also use both methods in the various partial solutions that may be involved in a respective application.

The step-by-step method consists of the successive decomposition of the experimental ensemble into simpler individual tests, applying theoretical results to each part of the test and combining the partial results to obtain the probabilities of the events to be measured.

The condensed method consists of treating the experimental ensemble as a whole unit (one single test), in which the events to be measured are decomposed according to the field of events attached to the respective experiment and the theoretical results are applied directly to the events to be measured.

Condensed methods are specific to the usage of combinatorics or classical probability schemes.

To see how the two methods work, let us consider a few very simple applications.

Exercises and problems

1) Two dice are rolled. Find the probability for the first die to show 3 and the second to show 5.

A. *Step-by-step method*
We consider the rolling of the two dice as two separate experiments (tests): rolling the first die and rolling the second.

We assume these tests are independent.

Denote by A – *the first die shows 3* and by B – *the second die shows 5.*

The two experiments are the same type (rolling the die) and have the same sample space, so we can assume that the two attached fields of events are identical.

Thus, event $A \cap B$ – *the first die shows 3 and the second die shows 5* make sense with respect to the intersection operation.

We have $P(A) = 1/6$, $P(B) = 1/6$ (we have applied the classical definition of probability for each part of the test) and, because A and B are independent events, we have

$$P(A \cap B) = P(A) \cdot P(B) = \frac{1}{6} \cdot \frac{1}{6} = 1/36 \text{ (this is the selected and}$$
applied formula).

B. *Condensed method*

We consider the rolling of the two dice as one single experiment.

The sample space attached to this experiment is the set of ordered pairs (a, b), with a, b $\in$ $\{1, 2, 3, 4, 5, 6\}$.

Their number is $6 \times 6 = 36$ and their occurrences are equally possible elementary events. Among them, a single pair is favorable for event E – *the first die shows 3 and the second die shows 5* to occur, namely the pair (3, 5).

We then have, according to the classical definition of probability, that $P(E) = 1/36$.

Observe that in the solution using the condensed method there was no longer a need for the additional formula of the probability of the intersection of two independent events, the classical definition of probability being sufficient.

2) A player participating in a card game with a 52-card deck is dealt two cards, with no cards in view at that moment. What is the probability of the player being dealt A♣ and 7♥? What is the probability of the player being dealt an *A* and a 7?

A. *Step-by-step method*

We consider the distribution of the two cards as two different experiments (the distribution of the first card and then the distribution of the second card, both performed by a dealer).

Denote the following events:

A – player is dealt an A♣ as the first card
B – player is dealt a 7♥ after the A♣ is dealt
C – player is dealt a 7♥ as first card
D – player is dealt an A♣ after the 7♥ is dealt

Events *A* and *C* belong to the field of events that is attached to first experiment, while *B* and *D* belong to the field of events attached to the second experiment.

Although this problem deals with two different probability fields, by identifying the events with the set of outcomes through which they occur, we see that these events are all part of the sample space of the first experiment; therefore, the intersection and union operations between them make sense.

We know that events A and B are independent, as are C and D, while A and C are incompatible.

Thus, events $A \cap B$ and $C \cap D$ are also incompatible.

$P(A) = 1/52$, $P(B) = 1/51$, $P(C) = 1/52$, $P(D) = 1/51$ (from the classical definition of probability).

The event to be measured is *player is dealt A♣ and 7♥* – $(A \cap B) \cup (C \cap D)$ and we have:

$$P\big((A \cap B) \cup (C \cap D)\big) = P(A \cap B) + P(C \cap D) =$$

$$P(A)P(B) + P(C)P(D) = \frac{1}{52} \cdot \frac{1}{51} + \frac{1}{52} \cdot \frac{1}{51} = 1/1326.$$

We have applied the formula of probability of intersection of two independent events and the formula of probability of union of two incompatible events.

B. *Condensed method*

We consider one experiment, namely, the distribution of two cards performed by dealer.

The set of elementary events attached to this experiment can be identified with the set of 2-size combinations from the 52, which has $C_{52}^2 = 1326$ elements.

Among them, a single combination is favorable for the occurrence of event to be measured *player is dealt A♣ and 7♥*, so its probability is $\dfrac{1}{C_{52}^2} = 1/1326$.

For the second question (the probability of a player being dealt any A and any 7):

A. *Step-by-step method*

We consider the distribution of the two cards as two different experiments (the distribution of the first card and then the distribution of the second card, both performed by a dealer).

Denote the following events:

A – player is dealt an A as the first card
B – player is dealt a 7 after the A is dealt
C – player is dealt a 7 as the first card
D – player is dealt an A after the 7 is dealt
We have here the same theoretical reasons as in first problem.
The probabilities for these events are as follows:
$P(A) = 4/52$, $P(B) = 4/51$, $P(C) = 4/52$, $P(D) = 4/51$ (from the classical definition of probability).
The event to be measured is *player is dealt an A and a 7 –* $(A \cap B) \cup (C \cap D)$ and we have:

$$P((A \cap B) \cup (C \cap D)) = P(A \cap B) + P(C \cap D) =$$

$$P(A)P(B) + P(C)P(D) = \frac{4}{52} \cdot \frac{4}{51} + \frac{4}{52} \cdot \frac{4}{51} = 16/1326.$$

B. *Condensed method*
We consider one experiment, namely, the distribution of two cards performed by a dealer.
The set of elementary events attached to this experiment can be identified with the set of 2-size combinations from the 52, which has $C_{52}^2 = 1326$ elements.

Among them, the combinations that are favorable for the occurrence of the event to be measured *player is dealt an A and a 7* are those of (A7) type, which number 4 x 4 = 16 (there are four aces and four sevens possible in the deck to be distributed), so its probability is $\dfrac{16}{C_{52}^2} = 16/1326$.

Observe that the solution using the condensed method is shorter and includes combinatorial calculus.

3) A coin is flipped five times. What is the probability of the coin showing heads exactly two times?

A. *Step-by-step method*
We consider the experiment of flipping the coin repeated five times.
The sample space is the same for all, namely, {H,T} (heads, tails).

Representing the repartition of outcomes of the five tests through 5-size arrangements (in which order counts), there are ten such arrangements that are favorable for the occurrence of the event to be measured, namely: (HHTTT), (HTHTT), (HTTHT), (HTTTH), (THHTT), (THTHT), (THTTH), (TTHHT), (TTHTH) and (TTTHH) (these are generated by the rule of each having only two H-symbols).

Each of these events has the probability

$\frac{1}{2} \cdot \frac{1}{2} \cdot \frac{1}{2} \cdot \frac{1}{2} \cdot \frac{1}{2} = \frac{1}{2^5} = \frac{1}{32}$, because it is an intersection of five

independent events (for example, (HHTTT) = *heads on first flip* and *heads on first flip and heads on second flip* and *heads on first flip and heads on second flip and tails on third flip* and *heads on first flip and heads on second flip and tails on third flip and tails on fourth flip* and *heads on first flip and head on second flip and tails on third flip and tails on fourth flip and tails on fifth flip*).

These ten events are mutually exclusive, so the probability of the event to be measured is the sum of their probabilities, or 10 x (1/32) = 10/32.

We can condense this entire calculation if we consider the repartition of outcomes of the five tests as 5-size combinations (in which order does not count) of outcomes, namely, the set of combinations of (abcde) type, with a, b, c, d, e $\in \{H, T\}$.

They total $2^5 = 32$, because each element can take two values.

Among them, those that are favorable for the event to be measured are those containing exactly 2 H-elements and their number is $C_5^2 = 10$. The probability is then 10/32.

This approach transforms the solving method into a condensed one.

B. *Condensed method*

Another condensed method for solving this problem is to use a classical probability scheme.

We can observe that a problem's hypothesis matches the Bernoulli scheme (of repeated experiments where the probability of occurrence of same event for a fixed number of times is requested).

We have only to apply the probability formula corresponding to this scheme, namely, $P_{m,n} = C_n^m p^m q^{n-m}$, by replacing the specific numerical values.

In this case we must calculate $P_{2,5} = C_5^2 \cdot \dfrac{1}{2^2} \cdot \dfrac{1}{2^3} = \dfrac{C_5^2}{2^5} = 10/32$.

Even from these few simple examples we can make the following conclusions about the differences between the two solving methods:
– The condensed method builds a solution that is shorter than the solution arrived at by the step-by-step method;
– The step-by-step method assumes simpler, but most often repeated arithmetic calculations, than the condensed method;
– The use of the step-by-step method requires greater attention in defining the involved events and in applying formulas than the condensed method;
– The use of condensed method generally assumes greater knowledge (of probability theory and combinatorial calculus) than the step-by-step method.

As we said earlier, an application can be solved in several manners by using either or both methods.

Choosing one method or another or a procedure that combines the two is an option that should take into account the application itself, as well as the solver's knowledge level, experience and mathematical calculus skills.

Although the use of step-by-step method is generally recommended to persons with minimal mathematical background or with no experience in combinatorial calculus, we recommend without reservation using the condensed method when possible because the step-by-step method is much more predisposed to observation, application and numerical calculus errors.

As an exercise, try to solve the following applications and focus on framing the problem and choosing the best solving method:

1) Two persons aleatory choose a number from 1 to 10 each. Find the probability of:

a) The first person choosing number 5 and the second choosing number 7;

b) Numbers 5 and 7 being chosen, no matter the order;

c) Two even numbers being chosen;

d) Both persons choosing the same number.

2) Three dice are simultaneously rolled. Find the probability for:

a) Three uneven numbers to be shown;

b) The number 4 to be shown on exactly two dice;

c) Three consecutive numbers to be shown (no matter the order).

3) Three cards are aleatory drawn from a 52-card deck. Calculate the probability of:

a) The three cards containing exactly one A;

b) The three cards containing a pair of 8's;

c) The three cards containing at least one clubs card (♣).

4) Two dice are rolled eight times. Calculate the probability for:

a) A pair (two identical numbers) to be shown on the first roll;

b) A pair to be shown exactly three times;

c) A pair to be shown at least four times;

5) Two persons are aleatory drawn from the crowd. Find the probability of the two persons not having the same birthday.

Selecting the formulas to use

After framing the problem and establishing the solving method, we must find the formulas to be applied in calculating the probabilities of the events to be measured.

These formulas can represent the basic properties of probability, theorems or results of the probability schemes previously studied, and they can be fully reviewed in the chapter on mathematics.

Probability calculus actually means the practical application of probability theory results in specific cases, including the numerical calculations here.

Although these formulas can cover the entire range of the finite and discrete applications and, theoretically, they might be mechanically applied through analogies, we do not recommend that anyone learning probability calculus try to reduce the whole mathematical chapter to this group of formulas for calculation purposes, because their correct application, as well as framing the problem, requires at least an ensemble vision of the theory.

Once the formulas to be used are selected, they can be applied by replacing the numerical values and the events specific to the problem and then performing the arithmetic or algebraic calculus.

We now present the complete list of these formulas, starting from Boolean operations.

The combinatorial calculus formulas are also included.

An identification number is attached to each formula.

We refer to these numbers in all applications that follow from here on in this chapter.

List of formulas to use

Boolean operations – applicable to sets of parts (corresponding to operators $\cup, \cap, {}^c$) and to events (corresponding to the operators *or, and, non)*:

(F1) $\begin{cases} A \cap (B \cup C) = (A \cap B) \cup (A \cap C) \\ A \cup (B \cap C) = (A \cup B) \cap (A \cup C) \end{cases}$ (distributivity)

(F2) $\begin{cases} (A \cup B)^c = A^c \cap B^c \\ (A \cap B)^c = A^c \cup B^c \end{cases}$ (De Morgan relations)

These are applied in problems where decomposing certain composed events is necessary.

Properties of probability

(F3) $P = m/n$ (the probability of an event is the ratio between the number of cases favorable for that event to occur and the number of equally possible cases) (the classical definition of probability)

(F4) $P(\Omega) = 1$ (probability of sure event)

(F5) $P(A \cup B) = P(A) + P(B)$, for any $A, B \in \Sigma$ with $A \cap B = \phi$.

(F6) $P\left(\bigcup_{i=1}^{n} A_i\right) = \sum_{i=1}^{n} P(A_i)$, for any finite family of mutually exclusive events $(A_i)_{i=1}^{n}$ (finite additivity in condition of incompatibility)

(F7) $P(A^c) = 1 = P(A)$ (probability of contrary event)

(F8) $P(\phi) = 0$ (probability of impossible event)

(F9) $P(A - B) = P(A) - P(A \cap B)$ (probability of difference of two events)

(F10) $P(A \cup B) = P(A) + P(B) - P(A \cap B)$ (general formula of probability of union of two events)

(F11) $P(A_1 \cup A_2 \cup \dots \cup A_n) = \sum_{i=1}^{n} P(A_i) - \sum_{j<i} P(A_i \cap A_j) +$

$\sum_{i<j<k} P(A_i \cap A_j \cap A_k) + \dots + (-1)^{n-1} P(A_1 \cap A_2 \cap \dots \cap A_n)$

(inclusion-exclusion principle or the general formula of probability of finite union of events)

This formula is a generalization of (F10) and is applied as follows:

We have n events and want to calculate the probability of their union. We consider successively all 1-size combinations, 2-size combinations and so on until n-size combinations of the n events (we have a single 1-size combination and a single n-size combination). We add the probabilities of unions of each group of same-size combinations. For the groups of combinations having an even number as a dimension, the total result is add then with minus (subtracted) and for those with an uneven number as a dimension the total result is add with plus (addition).

(F12) $P(A \cap B) = P(A) \cdot P(B)$, if events A and B are independent (the definition of independent events).

(F13) $P(A|B) = \dfrac{P(A \cap B)}{P(B)}$ (the definition of conditional probability)

(F14) $P(A) = \sum_{i \in I} P(A_i) P(A|A_i)$, if $(A_i)_{i \in I} \subset \Sigma$ is a complete system of events with $P(A_i) \neq 0, \forall i \in I$ (the formula of total probability)

(F15) $P(A_i|A) = \dfrac{P(A_i) \cdot P(A|A_i)}{\sum\limits_{i \in I} P(A_i) \cdot P(A|A_i)}$, if $(A_i)_{i \in I} \subset \Sigma$ is a complete system of events with $P(A_i) \neq 0, \forall i \in I$ (Bayes's theorem).

These are the basic formulas of probability calculus for finite and discrete cases.

Any step-by-step method of solving and most of the condensed methods can be reduced to successive direct applications of such formulas.

Statistical formulas

Let X be a discrete random variable that takes the values x_i, with probabilities $p_i, i \in I$.

(F16) $M(X) = \sum_{i \in I} x_i p_i$ (expected value or mean of X)

(F17) $\alpha_r(x) = M(X^r) = \sum_k x_k^r p_k$ (the r-th moment of X)

(F18) $\beta_r(X) = M(|X|^r) = \sum |x_k|^r p_k$ (the r-th absolute moment of X)

(F19) $\mu_r(X) = \alpha_r(X - M(X))$ (the centered r-th moment of X)

(F20) $D^2(X) = \sigma^2 = \mu_2(X)$ (dispersion or variance of X)

(F21) $D(X) = \sigma = \sqrt{\mu_2(X)}$ (standard deviation of X)

(F22) $D^2(X) = M(X^2) - [M(X)]^2$ (the relation between expected value and dispersion)

Formulas of the classical probability repartitions:

Bernoulli (binomial) scheme

n independent experiments are performed. In each an event A occurs with probability p and does not occur with probability $q = 1 - p$.

The probability for event A to occur exactly m times in the n experiments is:

(F23) $P_{m,n} = C_n^m p^m q^{n-m}$

(F24) $M(X) = np$ (expected value of a binomially obtained random variable)

(F25) $D^2(X) = npq$ (dispersion of a binomially obtained random variable).

Poisson scheme

n independent experiments are performed. In each an event A occurs with probability $p_i, (i = 1, 2, ..., n)$ and does not occur with probability $q_i = 1 - p_i$.

The probability for event A to occur exactly m times in the n experiments ($P_{m,n}$) is the coefficient of z^m from the product

(F26) $\varphi_n(z) = \prod_{i=1}^{n}(p_i z + q_i)$ (the generating function of

probabilities $P_{m,n}$)

The probability for A to occur at least m times in the n experiments is:

(F27) $P(C_m) = \sum_{k=m}^{n} P_{k,n}$

Polynomial scheme

An urn contains balls of colors $c_1, c_2, ..., c_s$, and p_i is the probability of drawing a ball of color $c_i, i = 1, ..., s$.

n draws of one ball each are performed, provided that the urn has the same content at each extraction.

The probability of α_i balls of color $c_i, i = 1, ..., s$ occurring on the performed draws is:

(F28) $P(n; \alpha_1, ..., \alpha_s) = \dfrac{n!}{\alpha_1!\alpha_2!...\alpha_s!} p_1^{\alpha_1}...p_s^{\alpha_s}$

The scheme of nonreturned ball

An urn has the following contents: a_1 balls of color c_1, a_2 balls of color $c_2, ..., a_s$ balls of color c_s.

n draws are performed without putting back the extracted ball (or just one draw of n balls).

The probability of occurrence of exactly α_k balls of color $c_k, k = 1, ..., s$ in the group of n extracted balls, where

$\alpha = (\alpha_1, ..., \alpha_s)$, $0 \leq \alpha_k \leq a_k$, $\sum_{k=1}^{s} \alpha_k = n$, is:

(F29) $P(n; \alpha_1, ..., \alpha_s) = \dfrac{C_{a_1}^{\alpha_1} C_{a_2}^{\alpha_2}...C_{a_s}^{\alpha_s}}{C_{a_1+...a_s}^{n}}$.

We now present a few solved applications and specify the formulas used through their identification number.

Exercises and problems

1) Two shooters are shooting one shot each at a target.
The probability of the first shooter hitting the target is 0.7 and the probability of the second hitting the target is 0.8. What is the probability of both shooters missing the target?

Answer:
Denoting by *A* the event *first shooter hits the target* and by *B* the event *second shooter hits the target,* the event to be measured is $(A \cup B)^c$.

According to (F2), we have $(A \cup B)^c = A^c \cap B^c$.

Events *A* and *B* are independent, so events A^c and B^c are independent.

Using (F12) and then (F7) twice, we get:
$$P\big((A \cup B)^c\big) = P\big(A^c \cap B^c\big) = P\big(A^c\big)P\big(B^c\big) = [1 - P(A)] \cdot [1 - P(B)].$$

By replacing the numerical values for P(A) and P(B), we obtain $0.3 \times 0.2 = 0.06 = 6\%$.

2) At a slot machine with three reels and seven symbols, find the probability of getting the same symbol on all reels in one spin.

Answer:
Let $S_1, S_2, S_3, S_4, S_5, S_6, S_7$ be the seven symbols and let us denote the following events:

A_i – symbol S_i occurs on the first reel;

B_i – symbol S_i occurs on the second reel;

C_i – symbol S_i occurs on the third reel;

The event to be measured is then $D = \bigcup_{i=1}^{7} (A_i \cap B_i \cap C_i)$.

Events $(A_i \cap B_i \cap C_i)$, $i = 1, ..., 7$ are mutually exclusive, so we have, according to (F6): $P(D) = \sum_{i=1}^{7} P(A_i \cap B_i \cap C_i)$.

Events A_i, B_i, C_i are independent in their totality for any *i*, so we have, according to (F13):

254

$$P(A_i \cap B_i \cap C_i) = P(A_i)P(B_i)P(C_i) = \frac{1}{7} \cdot \frac{1}{7} \cdot \frac{1}{7} = \frac{1}{7^3}, \forall i = 1, ..., 7.$$

Results $P(D) = 7 \cdot \dfrac{1}{7^3} = 1/49$.

This solution can be compressed through the following verbal reasoning: by fixing a certain reel and a certain symbol, the probability for that symbol to occur on that reel is 1/7. We have three reels spinning independently; therefore, the probability for that symbol to occur on all three reels is 1/7 x 1/7 x 1/7. We have seven such symbols, so the final probability is 7 x (1/7 x 1/7 x 1/7) = 1/49.

Exercise:
The same problem for a slot machine with four reels and eight symbols.

3) An urn contains nine black balls and fifteen white balls. Two balls are drawn from the urn.
What is the probability for the drawn balls to have same color?

Answer:

Solution 1 – step-by-step method
Denote the following events:
A – a white ball is drawn first, without putting it back;
B – a white ball is drawn second;
C – a black ball is drawn first, without putting it back;
D – a black ball is drawn second;
The event to be measured is E – *2 white balls or 2 black balls are drawn* and we have $E = (A \cap B) \cup (C \cap D)$.

Events $(A \cap B)$ and $(C \cap D)$ are incompatible.
Event B is conditioned by A and event D is conditioned by C. We have, according to (F14):
$$P(A \cap B) = P(A) \cdot P_A(B) = \frac{9}{24} \cdot \frac{8}{23};$$
$$P(C \cap D) = P(C) \cdot P_C(D) = \frac{15}{24} \cdot \frac{14}{23}.$$

By using (F5) and (F12), we have that
$$P(E) = P(A \cap B) + P(C \cap D) = 282/552 = 0.51086.$$

Solution 2 – condensed method

We consider the field of events in which the equally possible elementary events are the occurrences of 2-size combinations of balls from among the twenty-four balls. Their number is C_{24}^2.

The combinations that are favorable for the event to be measured E to occur are of the (WW) and (BB) type (W – white ball, B – black ball), for a total of $C_{15}^2 + C_9^2$.

According to (F3), we have that $P(E) = \dfrac{C_{15}^2 + C_9^2}{C_{24}^2} = 282/552$.

If the problem was about the probability of the two drawn balls having different colors, the favorable combinations would be of the (WB) and (BW) type, for a total of 2 x (9 x 15) = 270, so the probability would be $P(E) = \dfrac{270}{C_{24}^2} = 270/552 = 0.48913$.

Do the following check on the correctness of these calculations: while *two balls of the same color* and *two balls of different colors* are the only possibilities, their added probabilities should give 1, according to (F4). We have 282/552 + 270/552 = 1.

Exercise:

If an urn contains seven red balls and eight green balls and two balls are drawn, find the probability for the drawn balls to have:

a) the same color;
b) different colors.

4) Three candidates are present for an exam.

They can pass the exam if they know the subject from the drawn ticket well.

The first candidate learned 70 percent of the exam matter, the second 55 percent and the third 80 percent.

What is the probability for all three candidates to pass the exam?

Answer:
By making the approximations: the exam tickets uniformly cover the entire matter, a subject that belongs to the learned matter is exposed completely and correctly, the examiner is impartial, and there are no examination or communication errors, we are now in a case of a classical Poisson scheme with

$p_1 = 70/100$, $p_2 = 55/100$, $p_3 = 80/100$ and

$q_1 = 30/100$, $q_2 = 45/100$, $q_3 = 20/100$ and we must calculate $P_{3,3}$.

According to (F27), $P_{3,3}$ is the coefficient of z^3 from the

development of $\varphi_3(z) = \left(\dfrac{70}{100}z + \dfrac{30}{100}\right)\left(\dfrac{55}{100}z + \dfrac{45}{100}\right)\left(\dfrac{80}{100}z + \dfrac{20}{100}\right)$,

namely, $\dfrac{70}{100} \cdot \dfrac{55}{100} \cdot \dfrac{80}{100} = 0.308$.

Exercise:
In the same conditions, find the probability for exactly two candidates to pass the exam and the probability for at least one candidate to pass the exam.

5) You are participating in a Texas Hold'em Poker game with two opponents, you are dealt two hearts (♥) cards and the flop cards contain exactly two hearts cards. Calculate the probability for at least one of your opponents to hold two hearts cards.

Answer:

Solution 1:
We are in the hypothesis that you have five viewed cards (your hand and the three flop cards), from which four are hearts.

There remain forty-seven unseen cards, of which nine are hearts (13 – 4).

The nine unseen cards remaining are also called *outs*.

This is the initial information that generates the probability field and the final numerical result of probability.

The equally possible elementary events are the occurrences in your opponents' hands of the 2-size combinations of unseen cards.

Denoting event A – *the first opponent holds 2 hearts* and event

B – *the second opponent holds 2 hearts,* the event to be measured, *at least one of your opponents holds 2 hearts,* is $A \cup B$.

To calculate $P(A \cup B)$ by using (F10), we must first calculate $P(A)$, $P(B)$ and $P(A \cap B)$.

Because the same initial information is taken into account and the card distribution is aleatory, we have that $P(A) = P(B)$.

Then, we fix one of opponents (let this be the first), and make the calculus for this one, the result being the same for the other.

The equally possible elementary events that are favorable for event *A* to occur are the occurrences of combinations of (HH) type in the opponent's hand (H – hearts).

Their number is C_9^2 (all 2-size combinations from the nine cards of hearts left unseen).

The total number of possible combinations of cards the opponent can be dealt is C_{47}^2 (all 2-size combinations from the forty-seven cards left unseen).

According to (F3), we have that $P(A) = \dfrac{C_9^2}{C_{47}^2}$.

The same numerical value also stands for $P(B)$.

This would be the final answer if the problem were asked for a specific opponent to hold two hearts.

Let us now calculate $P(A \cap B)$:

Event $A \cap B$ occurs if each opponent holds an (HH) combination.

This means the occurrence of a double combination (HH)(HH) in the hands of the two opponents. Let us count these double combinations: (CC)(CC) = (CC) x (CC).

The first hand may be dealt C_9^2 combinations of cards (two hearts from the nine left). For each, the second hand may be dealt C_7^2 combinations of card (two hearts from the 9 – 2 = 7 left after two are counted in the first hand).

In total, we then have $C_9^2 \cdot C_7^2$ double combinations that are favorable for event $A \cap B$ to occur.

Let us now count the total number of possible double combinations the two hands can be dealt.

These are of the (xy)(zt) type, where x, y z and t are cards different from the five viewed.

$$(xy)(zt) = (xy) \times (zt)$$

We have C_{47}^2 possible combinations for the first hand (xy) and C_{45}^2 possible combinations for the second hand (zt), so we have a total of $C_{47}^2 \cdot C_{45}^2$ possible double combinations.

We now apply (F3) to find the probability of $A \cap B$ (in the probability field in which the equally possible elementary events are the occurrences of the double combinations in the hands of the two opponents): $P(A \cap B) = \dfrac{C_9^2 \cdot C_7^2}{C_{47}^2 \cdot C_{45}^2}$.

Returning to (F10) and replacing the numerical values found, we have that $P(A \cup B) = 2 \cdot \dfrac{C_9^2}{C_{47}^2} - \dfrac{C_9^2 \cdot C_7^2}{C_{47}^2 \cdot C_{45}^2} = \dfrac{C_9^2}{C_{47}^2}\left(2 - \dfrac{C_7^2}{C_{45}^2} \right)$.

Exercise:
Perform the combinatorial calculus and express the found probability as a percentage.

Solution 2:
Analyzing in greater detail the experiment involving card distribution to two opponents, we can observe that it corresponds to the nonreturned ball probability scheme for two tests and two colors of balls.

The urn is represented by the forty-seven unseen cards, the balls are represented by the 2-size combinations of cards from the forty-seven and the draw of a ball of a certain color is represented by the occurrence of an (HH) combination.

The distribution of the two 2-size combinations of cards is equivalent to the successive draw of two balls from the urn, without putting back the first drawn ball.

To apply (F30), let us observe the following analogies with the nonreturned ball probability scheme, for $n = 2$ (the number of tests) and $s = 2$ (the number of colors):

a_1 balls of color c_1 are equal in number to $C_9^2 = 36$ combinations of (HH) type;

a_2 balls of color c_2 are equal in number to $C_{47}^2 - C_9^2 = 1045$ combinations of non-(CC) type (that is, the combinations in which at least one card is not hearts).

With the denotations from formula (F30), the event to be measured is $A_{1,1} \cup A_{2,0}$ (one ball of color (CC) or two balls of color (CC) drawn as result of the two tests).

The two events are incompatible; therefore,

$P(A_{1,1} \cup A_{2,0}) = P(A_{1,1}) + P(A_{2,0})$, according to (F5).

By applying (F30) twice, we get:

$$P(A_{1,1}) = P(2; 1, 1) = \frac{C_{36}^1 \cdot C_{1045}^1}{C_{1081}^2} = \frac{36 \cdot 1045}{C_{1081}^2}$$

$$P(A_{2,0}) = P(2; 2, 0) = \frac{C_{36}^2 \cdot C_{1045}^0}{C_{1081}^2} = \frac{C_{36}^2}{C_{1081}^2}.$$

By adding the two results, we find the probability of the event to be measured is $\dfrac{36 \cdot 1045 + C_{36}^2}{C_{1081}^2}$.

Exercise:
Perform these calculations to their conclusion to ascertain whether we arrived at the same result as we achieved for solution 1.

Both methods of solving are of condensed type.

Although the second solution is shorter, we recommend that beginners use the first method to solve such applications because the application of the classical probability scheme assumes the correct identification of several equivalencies and analogies, and this action is predisposed to errors for a solver with no consistent mathematical practice.

In addition, the second solution cannot ignore the use of combinatorial calculus.

The calculus

The calculus is, in practice, the last stage of the algorithm of solving and cannot precede any other stage.

After the theoretical framework for a problem is established, selected formulas can be applied by replacing the numerical values from the hypothesis or deducted values in them.

Then comes the work on the numerical calculations to obtain the final result.

Probability calculus actually means to apply selected formulas and obtain a final result by replacing the numerical values within formulas and working the numerical calculations.

The numerical calculus itself can be strictly arithmetic as well as combinatorial.

As we mentioned earlier, users of this guide must have a good command of arithmetic and basic algebraic calculus before they can complete the calculations: operations with real numbers, order of operations, operations with fractions, reductions, expanding the brackets, multiplication of expressions within brackets, raising an expression to a power, factoring out, and the like.

The combinatorial calculus is present in the solutions of numerous probability applications, and the user must have a good command of it as well.

In most probability problems in which combinatorial calculus is involved, the application of formulas actually reverts to counting permutations, combinations or arrangements and performing the arithmetic calculations between them.

That is why we recommend that readers to study the chapter titled *Combinatorics* and solve as many combinatorial calculus exercises and problems as possible before approaching the probability applications.

Knowing the general formulas for permutations, arrangements and combinations is very important but not sufficient: the practice has a decisive role in framing a combinatorial problem correctly and in the proper application of formulas.

The background practice is also useful for the arithmetic calculus because it also develops the ability to observe, which is necessary to avoid errors, to simplify calculations and to make proper approximations.

Numerical approximations may be used when the application of some long and heavy formulas requires a hard and over-long calculus, provided these approximations do not change the real result too much.

The approximations are performed by removing from formulas those terms and expressions having a numerical value that is insignificant to the conditions and requests of an application.

Odds and probability

The final numerical result of a probability application can be expressed in three ways: as a fraction, as a percentage or as odds.

Probability, as a subunit real number, is a fraction.

Converting it into a percentage is a simple matter of arithmetic division and denotation.

For example, to express the probability 1/8 as a percentage, we must divide 1 : 8, to get 0.125, which is denoted as the percentage 12.5%.

The last action consists in fact of moving the decimal point (or coma) toward the right hand by passing two decimal places (figures) and adding the % sign.

Conversely, expressing a percentage as fraction assumes writing the number obtained after removing the decimal point (or coma) and the % sign as a numerator and the power of 10 with the exponent being the sum of the number of decimal places of the initial number and 2 as the denominator.

Let us write the number 12.5% as a fraction:

We have one decimal place. Removing the decimal point (or coma) and %: 125. Calculating: $10^{1+2} = 1000$. Write the result as a fraction: 125/1000. If we continue to reduce this fraction, we obtain 1/8.

Let us write the number 0.153% as a fraction:

We have three decimal places. Removing the decimal point (or coma) and %: 0153. Calculating: $10^{3+2} = 100000$. Write the found fraction: 153/100000.

The *odds* represent a way of expressing the probability of an event reported to the probability of the contrary event.

The following formula is used to convert a probability into odds:

odds = probability / (1 − probability)

For example, a 1/3 probability of event A is converted into odds as follows:

odds of $A = (1/3)/(1 − 1/3) = (1/3)/(2/3) = 1/2$

This is denoted by 2 : 1 and reads *the chances of A to occur are 2 against 1,* meaning *there are two chances in three for* A^c *to occur and one chance in three for* A *to occur.*

The expression of probability through odds usually holds only natural numbers, so we must make an approximation if necessary.

For example, if the above formula returns $3/7 = 1/2.333...$, we can write 3:1 as odds.

Conversely, to convert odds into a probability, the following formula is used:

probability = odds / (1 + odds)

For example, a 3:2 odds is converted into probability as follows:
$(2/3) / (1 + 2/3) = (2/3) / (5/3) = 2/5$ or 40% as percentage.

Even the words *probability* and *odds* have different definitions attached, they express the same quantitative and qualitative content, so their undifferentiated usage along with the associated figures is not quite an error, but rather a deviation from consistent terminology.

We also present a few solved applications in this section, for which we included the development and finalization of the arithmetic and combinatorial calculations.

Exercises and problems

1) You are participating in a 52-card deck classical poker game with three opponents and you are dealt (A♥ A♣ J♣ 7♣ 2♣).

What is the probability for at least one opponent to hold an ace (calculated in the moment before the second card distribution)?

What is the probability that you receive a card of clubs if you discard A♥?

What is the probability that you receive at least one ace at the second card distribution if you discard J♣ 7♣ 2♣?

Answer:

An ace means at least one ace, or one or two aces (taking into account that you already hold two of the four aces).

Denote by A the event to be measured: *at least one opponent holds at least one ace*.

If we denote by A_i the events *opponent number i holds at least one ace*, i = 1, 2, 3, we have that $A = A_1 \cup A_2 \cup A_3$.

Obviously, events A_i have the same probability in the probability field we operate within (in which the equally possible elementary events are the occurrences in the opponents' hands of 5-size combinations of cards from the $52 - 5 = 47$ cards left unseen).

Let us calculate first $P(A_i)$ for an arbitrary fixed opponent i.

The combinations favorable for event A_i are those of the (Axyzt) type, where x, y, z and t take any value different from the five cards seen (including ace).

To count these combinations, we split them into two distinct groups: those containing exactly one A (ace) and those containing exactly two A (we have no combinations containing more than two aces, because you hold two of the four aces available).

We denote these groups by (Axyzt) and (AAxyz), with $x, y, z, t \neq A$. Obviously, there are no common combinations of the two groups, so we can sum the numbers of combinations corresponding to each group.

Let us calculate the number of combinations (Axyzt) by partitioning:

(Axyzt) = (A)(xyzt)

264

Two aces are in play, so the number of 1-size combinations (A) is two.

To calculate the number of 4-size combinations (xyzt) that correspond to each combination (A), we first count the number of cards representing the set from which these combinations are generated: from the initial 52 cards, we subtract the five from our own hand, we subtract one more for (A) and another one for the remaining ace (because x, y, z and t are different from A).

We then have $52 - 5 - 1 - 1 = 45$, so we have C_{45}^4 combinations (xyzt).

We then have $2C_{45}^4$ combinations (Axyzt).

Let us now calculate the number of combinations (AAxyz), still by partitioning:

(AAxyz) = (AA)(xyz)

The number of combinations (AA) is $C_2^2 = 1$ (the single combination of two aces left).

The 3-size combinations (xyz) are generated from a set that has $52 - 5 - 2 = 45$ elements (from the initial 52 cards we subtracted the five from our own hand and the two aces from the combination (AA); there are no other aces in play), so their number is C_{45}^3.

The result is that the number of combinations (AAxyz) is $1 \cdot C_{45}^3 = C_{45}^3$.

By summing the two partial results, we get the number of combinations containing at least one A. This number is $2C_{45}^4 + C_{45}^3$.

This is the number of combinations that are favorable for the event to be measured A_i.

The number of all possible combinations an opponent can receive is C_{47}^5 (five cards from the $52 - 5 = 47$ remaining in play).

Now we can apply (F3) to find the probability of A_i:

$$P(A_i) = \frac{2C_{45}^4 + C_{45}^3}{C_{47}^5}, \text{ for every i} = 1, 2, 3.$$

Let us return to the count of combinations that are favorable for A_i with a parenthesis:

The combinations are split into two groups to avoid a frequent calculation error; namely, the multiple count of some combinations.

Several persons leave those combinations in the form (Axyzt), with x, y, z and t taking any value (less the five cards from their own hand) and count them in the following manner:

(Axyzt) = (A)(xyzt)

Combinations (A) number 2 (two aces left) and the combinations of the (xyzt) type number $C_{52-5-1}^4 = C_{46}^4$ (the number of cards from which the 4-size combinations are generated is obtained by subtracting from 52 the five cards from their own hand and the one ace counted in the first combination (A)).

The searched number is then $2C_{46}^4$.

Well, this ratiocination is wrong because it allows some combinations to be counted twice:

Because we allowed elements x, y, z and t to also take the value A, a combination of the (AAyzt) type is counted twice through this ratiocination—once as (A♠ A♦ yzt) and once as (A♦ A♠ yzt)—when, in fact, they represent one and the same combination.

Exercise: Do the combinatorial calculus to the end in both ratiocinations (the correct one and the wrong one) and check whether the result of the wrong one is larger than the correct one.

Bear in mind this error of partitioning combinations.

The rule of a correct calculus in problems in which we have conditions of the type *at least* n *elements of a certain type in a combination* is splitting the combinations to count into two or more incompatible groups, each containing exactly 1, 2, ..., respectively n elements of that type.

Let us now return to the initial problem. We calculated $P(A_i)$. To calculate $P(A)$, we apply (F11).

We have three events A_i and we must calculate the probabilities of the intersections of two and three events from them:

The intersections of two events:

We have $C_3^2 = 3$ such intersections. Each has the same probability.

We fix two opponents; for example, opponents 1 and 2.

The event $A_1 \cap A_2$ is *opponent 1 and opponent 2 have at least one ace each.*

The number of all possibilities of card distributions for the two opponents is given by the number of double combinations (xyztv)(mpqrs) that have all elements mutually different and different from the five cards shown. This number is $C_{47}^5 C_{42}^5$ (we have $C_{52-5}^5 = C_{47}^5$ combinations (xyztv) and for each combination (xyztv) exist $C_{52-5-5}^5 = C_{42}^5$ combinations (mpqrs), and the results must be multiplied).

Let us now count the double combinations that are favorable for event $A_1 \cap A_2$, where we can apply (F3).

These are of the (Ayztv)(Apqrs) type, with y, z, t, v, p, q, r, s different from A (there are no uncounted aces).

(Ayztv)(Apqrs) = (A)(yztv)(A)(pqrs)

We have two combinations (A) (two aces left in play);

We have C_{45}^4 combinations (yztv) (the 4-size combinations from 52 cards, from which we subtract the five cards shown, one ace counted in the previous combination and one ace left uncounted: $52 - 5 - 1 - 1 = 45$);

We have one combination (A) (one ace left uncounted);

We have C_{41}^4 combinations (pqrs) (the 4-size combinations from 52 cards, from which we subtract the five cards shown and the six cards from previous combinations; no ace remains uncounted: $52 - 5 - 6 = 41$).

By multiplying the partial results, we obtain $2C_{45}^4 + C_{41}^4$.

We now apply (F3) and find: $P(A_1 \cap A_2) = \dfrac{2C_{45}^4 C_{41}^4}{C_{47}^5 C_{42}^5}$.

The intersections of three events:

We have $C_3^3 = 1$, or a single intersection of three events.

A triple combination that is favorable for event $A_1 \cap A_2 \cap A_3$ should have the form (Axyzt)(Avpqr)(Abcde), which is impossible, because there are only two aces left in play and any such triple combination must contain three aces.

We then have $P(A_1 \cap A_2 \cap A_3) = 0$.

We now apply (F11) to find the searched probability:

$$P(A) = [P(A_1) + P(A_2) + P(A_3)] -$$
$$- [P(A_1 \cap A_2) + P(A_1 \cap A_3) + P(A_2 \cap A_3)] + P(A_1 \cap A_2 \cap A_3).$$

By replacing the deducted numerical values, we obtain:

$$P(A) = 3 \cdot \frac{2C_{45}^4 + C_{45}^3}{C_{47}^5} - 3 \cdot \frac{2C_{45}^4 C_{41}^4}{C_{47}^5 C_{42}^5} - 0 = \frac{3}{C_{47}^5} \left(2C_{45}^4 + C_{45}^3 - \frac{2C_{45}^4 C_{41}^4}{C_{42}^5} \right).$$

When dealing with combinations of large numbers, it is useful not to unfold the combinatorial calculus until the end, to allow for factoring out and eventual reductions.

This type of problem is seen frequently in applications corresponding to card games in which the requests are the probabilities of various events regarding opponents.

As you may have noticed, the problem is reduced to counting combinations and to the direct application of formula (F11).

Let us now answer the second question of the problem.

The event to be measured is B – *you receive a clubs at the second card distribution.*

You already hold four clubs, so there are still 13 – 4 = 9 clubs in play from a total of 47 unseen cards.

According to (F3), probability of *B* is P(B) = 9/47.

Let us now answer the last question of the problem.

The event to be measured is C – *you receive at least one* A *at the second card distribution.*

The equally possible elementary events are the occurrences of 3-size combinations from the 47 unseen cards. They total C_{47}^3.

The combinations that are favorable for event *C* are those containing at least one ace. Let us count them by splitting them into two distinct groups:

(Axy), with x, y ≠ A (those containing exactly one *A*), and

(AAx), with x ≠ A (those containing exactly two *A*).

In the first group: (Axy) = (A)(xy); we have two combinations (A) and C_{45}^2 combinations (xy) (two cards from 52, from which we subtract the five cards shown, the one ace counted in the previous combination and the one ace left uncounted: $52 - 5 - 1 - 1 = 45$); the number of combinations (Axy) is then $2C_{45}^2$.

In the second group: (AAx) = (AA)(x); we have one combination (AA) (only two aces are still in play) and $52 - 5 - 2 = 45$ combinations (cards) (x); the number of combinations (AAx) is then $1 \times 45 = 45$.

By summing the two partial results, we find $2C_{45}^2 + 45$ combinations that are favorable for event C from C_{47}^3 possible; therefore, we have $P(C) = \dfrac{2C_{45}^2 + 45}{C_{47}^3}$.

Observe that a gaming decision can be made on the basis of the three calculated probabilities at each gaming stage.

Exercise: Do the calculation at the end and compare the values of the requested probabilities. Write the found results as a fraction, a percentage and as odds.

2) You buy a ticket (one variant) for the 6/49 lottery. What is the probability of exactly three numbers printed on your ticket being drawn?

Answer:

Solution 1:
Let (abcdef) be the played variant (a, b, c, d, e, f are distinct natural numbers from 1 to 49).
Let us denote the events:
A_{abc} – exactly three numbers from your ticket are drawn: the first, the second and the third, namely, a, b, c;
A_{abd} – exactly three numbers from your ticket are drawn: the first, the second and the fourth, namely, a, b, d; and so on;
A_{def} – exactly three numbers from your ticket are drawn: the fourth, the fifth and the sixth, namely, d, e, f.

We have $C_6^3 = \dfrac{4 \cdot 5 \cdot 6}{2 \cdot 3} = 20$ such events A_{ijk} and the event to be

measured is their union: $\displaystyle\bigcup_{\substack{i,j,k \in \{a,b,c,d,e,f\} \\ i<j<k}} A_{ijk}$.

The total number of possible combinations that can be drawn is C_{49}^6 .

The probability of an event A_{ijk} can be calculated very simply: the combinations that are favorable to event A_{ijk} are of the form $(ijkxyz)$, with x, y, z distinct and different from a, b, c, d, e.

$(ijkxyz) = (ijk)(xyz)$

Their number is $1 \cdot C_{49-6}^3 = C_{43}^3$ (one combination (ijk) and x, y, z may take $49 - 6 = 43$ values).

According to (F3), we then have: $P(A_{ijk}) = \dfrac{C_{43}^3}{C_{49}^6}$, for any i, j, k.

Observe that events A_{ijk} are mutually exclusive; therefore, we are

allowed to apply (F6): $\displaystyle P\left(\bigcup_{\substack{i,j,k \in \{a,b,c,d,e,f\} \\ i<j<k}} A_{ijk}\right) = 20 \cdot P(A_{ijk}) = \dfrac{20 C_{43}^3}{C_{49}^6}$.

Exercise: Do the complete calculation and express the probability as a percentage.

Solution 2:
This experiment corresponds to the scheme of the nonreturned ball, with the following equivalencies:
– draw of a ball = draw of a number; the experiment is repeated for $n = 6$ times, without putting back the ball;
– we have two colors ($s = 2$): the numbers (balls) from the ticket (a, b, c, d, e, f) represent the first color, c_1 , and the remaining $49 - 6 = 43$ balls have the second color, c_2 ;
– we have $a_1 = 6$ balls of color c_1 and $a_2 = 43$ balls of color c_2 ;

The problem asks for the probability of $\alpha_1 = 3$ balls of color c_1 and $\alpha_2 = 3$ balls of color c_2 being drawn.

By applying (F30), we find the searched probability to be

$$P(6; 3, 3) = \frac{C_6^3 C_{43}^3}{C_{49}^6}.$$

3) Four dice are rolled independently. Find the probability of getting a pair, three identical numbers or four identical numbers (that is, at least two identical numbers).

Answer:

Solution 1:

The equally possible elementary events are the occurrences of the various 4-size combinations (xyzt) on the four dice (x, y, z, t are natural numbers from 1 to 6).

The total number of these combinations is $6 \times 6 \times 6 \times 6 = 6^4$.

Let us denote the events:

A_N – occurrence of a combination (NNxy), $x, y \neq N$, $N = 1, ..., 6$

B_N – occurrence of a combination (NNNx), $x \neq N$, $N = 1, ..., 6$

C_N – occurrence of a combination (NNNN), $N = 1, ..., 6$.

If D_N – occurrence of a combination (NNxy), x, y any, then

$$D_N = A_N \cup B_N \cup C_N \text{ and the event to be measured is } \bigcup_{N=1}^{6} D_N.$$

For a fixed N, let us calculate $P(A_N)$.

We must count the combinations that are favorable to A_N:

(NNxy) = (N)(N)(x)(y) = $1 \times 1 \times 5 \times 5 = 5^2$ (N takes 1 value and x, y take $6 - 1 = 5$ values each).

We then have $P(A_N) = \dfrac{5^2}{6^4}$, for any N (according to (F3)).

The combinations that are favorable to B_N:

(NNNx) = (N)(N)(N)(x) = $1 \times 1 \times 1 \times 5 = 5$ (N takes 1 value and x takes $6 - 1 = 5$ values).

We then have $P(B_N) = \dfrac{5}{6^4}$.

271

The combinations that are favorable to C_N:

$(NNNN) = (N)(N)(N)(N) = 1 \times 1 \times 1 \times 1 = 1$.

We then have $P(C_N) = \dfrac{1}{6^4}$.

The events A_N, B_N, C_N are mutually exclusive for any N; therefore, according to (F6), we have:

$$P(D_N) = P(A_N) + P(B_N) + P(C_N) = \frac{5^2}{6^4} + \frac{5}{6^4} + \frac{1}{6^4} = \frac{31}{6^4}.$$

To calculate $P\left(\bigcup_{N=1}^{6} D_N\right)$, we apply (F11).

We calculate the probabilities of $D_1, ..., D_6$ and then calculate the probabilities of the intersections of these sets (events).

Let us take, for example, $D_1 \cap D_2$: an event belonging to this intersection is the occurrence of a combination (11xy) and of a combination (22zt); that is, the occurrence of combination (1122).

Therefore, this intersection has a single element. We can write this schematically: $(11xy) \cap (22zt) = (1122)$.

The ratiocination is valid for any intersection of two sets D_N, so we have $P(D_i \cap D_j) = \dfrac{1}{6^4}$ for any $i \neq j$ (one favorable combination from 6^4 possible).

In total, we have C_6^2 intersections of two sets.

Any intersection of three or more sets D_N is empty.

For example, if it is not empty, the intersection $(11xy) \cap (22zt) \cap (33uv)$ should contain combinations holding 112233, and that is impossible (a combination has only four elements).

Obviously, the intersections of four or more sets will also be empty.

We can now apply (F11):

$$P\left(\bigcup_{N=1}^{6} D_N\right) = \sum_{N=1}^{6} P(D_N) - \sum_{\substack{i,j \in \{1,2,3,4,5,6\} \\ i<j}} P(D_i \cap D_j) + 0 = \frac{6 \cdot 31}{6^4} - \frac{C_6^2}{6^4} =$$

$$= \frac{186 - 15}{6^4} = \frac{171}{1296} = 0.13194 = 13.194\%.$$

Solution 2:

We apply the Bernoulli scheme three times for n = 4, p = 1/6, q = 5/6: for m = 2, m = 3 and m = 4. According to (F27), we have

$$P_{2,4} = C_4^2 \left(\frac{1}{6}\right)^2 \left(\frac{5}{6}\right)^2 = \frac{6 \cdot 5^2}{6^4} = \frac{150}{6^4}$$

$$P_{3,4} = C_4^3 \left(\frac{1}{6}\right)^3 \left(\frac{5}{6}\right) = \frac{4 \cdot 5}{6^4} = \frac{20}{6^4}$$

$$P_{4,4} = C_4^4 \left(\frac{1}{6}\right)^4 = \frac{1}{6^4}.$$

By summing the results, we get:

$$P\left(\bigcup_{N=1}^{6} D_N\right) = P_{2,4} + P_{3,4} + P_{4,4} = \frac{171}{6^4}.$$

4) Twelve competitors from three towns (four competitors from each town) are participating in a race. Find the probability for at least two competitors from the first four ranked to be from the same town.

Answer:

The number of possible ways the competitors can fill the four positions is C_{12}^4 (because we are interested in the cumulative number of competitors of a certain type from the first four positions, the order does not count).

By approximating: all competitors are equally trained and have the same physical strength, their origin does not influence their performance, none of them is cheating, etc., we can consider the occurrences of the various 4-size combinations from the twelve competitors as equally possible elementary events.

Denote by a a competitor from the first town, by b a competitor from the second town and by c a competitor from the third town.

Denote the events:

A – occurrence of a combination (aaxy), with any x, y;

B – occurrence of a combination (bbxy), with any x, y;

C – occurrence of a combination (ccxy), with any x, y.

The event to be measured is $A \cup B \cup C$.

Obviously, events A, B, C are not mutually exclusive (the first four positions can be filled by four competitors from two different towns, two competitors from each town).

Let us calculate $P(A)$. To obtain a correct count of combinations that are favorable to A, we split them into three incompatible groups, as follows:

(aaxy), with x, y different from a;

(aaax), with x different from a;

(aaaa).

$(aaxy) = (aa)(xy) = C_4^2 \cdot C_{12-2-2}^2 = C_4^2 C_4^8$ (we have four elements having value a, so C_4^2 combinations (aa), then for any combination (aa) twelve elements remain, from which we subtract the two elements from (aa) and the two elements of value a left from the initial four);

$(aaax) = (aaa)(x) = C_4^3(12-3-1) = 8C_4^3$ (we have four elements having value a, so C_4^3 combinations (aaa), then for any combination (aaa) twelve elements remain, from which we subtract the three elements from (aaa) and another one of value a left from the initial four);

$(aaaa) = C_4^4 = 1$.

If we add these partial results (there are no combinations that are common to the three groups or to any two of them), we obtain:

$$C_4^2 C_8^2 + 8C_4^3 + 1 = 6 \cdot \frac{7 \cdot 8}{2} + 8 \cdot 4 + 1 = 3 \cdot 7 \cdot 8 + 32 + 1 = 201$$

combinations that are favorable to event A.

According to (F3), we have, $P(A) = \dfrac{201}{C_{12}^4}$.

Events B and C have the same probability, so

$$P(A) = P(B) = P(C) = \frac{201}{C_{12}^4}.$$

To apply (F11), we first study the intersections of sets A, B and C.

The intersections of two sets:

We have $C_3^2 = 3$ intersections of two sets, each intersection having the same probability.

For the intersection $A \cap B$, $(aaxy) \cap (bbzt) = (aabb)$.

The number of these combinations is $(aabb) = (aa)(bb) = C_4^2 C_4^2 = 36$.

We then have $P(A \cap B) = \dfrac{36}{C_{12}^4}$ and this value is also valid for the other two intersections.

The intersections of three sets:

The intersection $A \cap B \cap C$ is empty.

If a common combination of the three sets existed, it should contain *aabbcc*, which is impossible (the combination has only four elements).

Therefore, $P(A \cap B \cap C) = 0$.

We now apply (F11):

$$P(A \cup B \cup C) = 3P(A) - 3P(A \cap B) + 0 = 3 \cdot \frac{201}{C_{12}^4} - 3 \cdot \frac{36}{C_{12}^4} = 165 \cdot \frac{3}{C_{12}^4}$$

Exercise: Do the complete calculus and write the result as a percentage.

5) You are participating in a Texas Hold'em Poker game with nine opponents and you were dealt a pair of 9's.

Calculate the probability for at least one opponent to hold a higher pair (before the flop).

Answer:

The pairs higher than (99) are of the type (10 10), (JJ), (QQ), (KK) and (AA).

We first calculate the probability for a specific opponent (a fixed one) to hold one of these pairs. Let A be this event.

The combinations that are favorable to A are those listed above, a total of $6C_4^2 = 36$ (six types of pairs, each having C_4^2 combinations).

The total number of combinations an opponent can hold is $C_{50}^2 = 1225$ (the 2-size combinations of cards from 52 minus the two cards shown).

275

We then have $P(A) = 36/1225$ and this value is valid for any opponent.

We have nine opponents. Denoting by A_i the event *opponent number i holds a pair that is higher than (99)*, $i = 1, ..., 9$, we see that the event to be measured is $B = \bigcup_{i=1}^{9} A_i$.

To calculate the probability of B, we apply (F11).

The intersection of two events A_i *:*

We have $C_9^2 = 36$ intersections of two events.

Such an intersection contains double combinations of the type (CC)(DD), with C and D taking the values 10, J, Q, K or A (five values).

Their number is $5C_4^2(5C_4^2 - 1) = 5 \cdot 6(5 \cdot 6 - 1) = 30 \cdot 29$ (the number of combinations (CC) is $5C_4^2$, because C can take five values and for each value there are four cards in play; for each combination (CC), there are $5C_4^2$ combinations (DD) minus one; namely, the previous combination (CC)).

The number of all possible double combinations (xy)(zt) the two opponents can hold is $C_{50}^2 C_{48}^2$ (the first opponent can receive any two cards from the 50 unseen cards and for each combination (xy) the second opponent can receive any two cards from the 50, minus the two cards from first opponent's hand).

The probability of an intersection of two events $A_i \cap A_j$ is then

$$P(A_i \cap A_j) = \frac{30 \cdot 29}{230300}.$$

The intersection of three events A_i *:*

We have $C_9^3 = 84$ intersections of three events.

Such an intersection contains triple combinations of the type (CC)(DD)(EE), with C, D and E taking the values 10, J, Q, K or A (five values).

Their number is
$$5C_4^2(5C_4^2 - 1)(5C_4^2 - 2) = 5 \cdot 6(5 \cdot 6 - 1)(5 \cdot 6 - 2) = 30 \cdot 29 \cdot 28$$

(the number of combinations (CC) is $5C_4^2$, because C can take five values and for each value there are four cards in play; for each combination (CC), there are $5C_4^2$ combinations (DD) minus one combination, namely, the previous combination (CC); for each combination (DD) there are $5C_4^2$ combinations (EE) minus the two previous combinations).

The number of all possible triple combinations (xy)(zt)(uv) the three opponents can hold is $C_{50}^2 C_{48}^2 C_{46}^2 = 238360500$ (the first opponent can receive any two cards from the 50 unseen cards; for each combination (xy) the second opponent can receive any two cards from the 50 minus the two cards from first opponent's hand; for each combination (zt) the third opponent can receive any two cards from the 50 minus the four from first two opponents' hands).

The probability of the intersection of three events $A_i \cap A_j \cap A_k$ is then $P(A_i \cap A_j \cap A_k) = \dfrac{30 \cdot 29 \cdot 28}{238360500}$.

Through a similar ratiocination, for the intersection of four events (a total of $C_9^4 = 126$), we find

$$P(A_i \cap A_j \cap A_k \cap A_r) = \frac{30 \cdot 29 \cdot 28 \cdot 27}{225489033000}.$$

This value is approximately 0.0000029. Multiplied by 126 (as it appears when formula (F11) is applied), it is approximately 0.000367.

If we continued the calculus of probabilities of intersections of 5, 6, 7, 8 and 9 events, we would obtain lower and lower numerical results, because the product from the numerator increases faster than the product from denominator.

The calculus becomes cumbersome because we are now dealing with very large numbers.

In this situation, we can do a numerical approximation by removing from formula (F11) all the terms containing the probabilities of the intersection of five events or more.

This approximation is justified not only by the very low value of the terms removed, but also by the fact that their signs alternate with plus and minus within formula (F11).

By applying (F11) and writing the terms until the probabilities of the intersection of four events are all included, we obtain:

$$P(B) = \frac{9 \cdot 36}{1225} - \frac{36 \cdot 30 \cdot 29}{230300} + \frac{84 \cdot 30 \cdot 29 \cdot 28}{238360500} - \frac{126 \cdot 30 \cdot 29 \cdot 28 \cdot 27}{225489033000} + \dots$$

Exercise: Do the complete calculation with the help of a calculator and write the found probability as a percentage.

PROBABILITY CALCULUS APPLICATIONS

Because skills in probability calculus and in correctly applying the theoretical results are acquired only through exercise, this chapter contains a collection of solved and unsolved applications that cover most of the range of classical probability problems.

The reader who has studied the *Beginner's Calculus Guide* can practice the theoretically acquired skills by studying the solutions to the classical problems and solving as many applications as possible.

We also recommend that readers who want to improve their application and probability calculus skills not limit themselves to the two chapters of applications in this book, but work on other specific problem books, too.

Solved applications

1) Determine $\mathcal{P}(\Omega)$ if:

a) $\Omega = \{1, 2\}$;

b) $\Omega = \{A\}$;

c) $\Omega = \{\{0, 1\}, \{2, 3\}, \{4\}\}$.

Solution:

a) $\mathcal{P}(\Omega) = \{\phi, \{1\}, \{2\}, \{1, 2\}\}$;

b) $\mathcal{P}(\Omega) = \{\phi, \{A\}\}$;

c) $\mathcal{P}(\Omega) =$
$\{\phi, \{\{0, 1\}\}, \{\{2, 3\}\}, \{\{4\}\}, \{\{0, 1\}, \{2, 3\}\}, \{\{0, 1\}, \{4\}\}, \{\{2, 3\}, \{4\}\}, \Omega\}$;

2) Write the sample space for the following experiments:

a) draw of a ball from an urn containing seven balls;

b) draw of two balls from two urns (one ball from each), the first containing three green balls and the second two red balls;

c) draw of a card from a 24-card deck (from the 9 card upward);

d) rolling two dice;

e) choosing three numbers from the numbers 1, 2, 3, 4, 5;

f) choosing seven letters from the letters a, b, c, d, e, f, g, h;

Solution:

a) By numbering the balls, we have:

$\Omega = \{$ball 1, ball 2, ball 3, ball 4, ball 5, ball 6, ball 7$\}$ or, equivalent,

$\Omega = \{1, 2, 3, 4, 5, 6, 7\}$.

b) By numbering the balls from the two urns and denoting by g a green ball and by r a red ball, the sample space is the following set of ordered pairs:

$\Omega = \{(1g, 1r), (1g, 2r), (2g, 1r), (2g, 2r), (3g, 1r), (3g, 2r)\}$.

c) $\Omega = \{9\spadesuit, 9\clubsuit, 9\heartsuit, 9\diamondsuit, 10\spadesuit, 10\clubsuit, 10\heartsuit, 10\diamondsuit, J\spadesuit, J\clubsuit, J\heartsuit, J\diamondsuit, Q\spadesuit, Q\clubsuit, Q\heartsuit, Q\diamondsuit, K\spadesuit, K\clubsuit, K\heartsuit, K, A\spadesuit, A\clubsuit, A\heartsuit, A\diamondsuit\}$.

d) The sample space is the following set of ordered pairs:

$\Omega = \{(1, 1), (1, 2), (1, 3), (1, 4), (1, 5), (1, 6), (2, 1), (2, 2), (2, 3),$
$(2, 4), (2, 5), (2, 6), (3, 1), (3, 2), (3, 3), (3, 4), (3, 5), (3, 6), (4,$
$1), (4, 2), (4, 3), (4, 4), (4, 5), (4, 6), (5, 1), (5, 2), (5, 3), (5, 4), (5,$
$5), (5, 6), (6, 1), (6, 2), (6, 3), (6, 4), (6, 5), (6, 6)\}$.

e) The sample space is the set of all 3-size combinations of numbers from the 5 given:

$\Omega = \{(1, 2, 3), (1, 2, 4), (1, 2, 5), (1, 3, 4), (1, 3, 5), (1, 4, 5),$
$(2, 3, 4), (2, 3, 5), (2, 4, 5), (3, 4, 5)\}$.

f) The sample space is the set of all 7-size combinations of letters from the 8 given:

$\Omega = \{(a, b, c, d, e, f, g), (a, b, c, d, e, f, h), (a, b, c, d, e, g, h), (a,$
b, c, d, f, g, h), (a, b, c, e, f, g, h), (a, b, d, e, f, g, h), (a, c, d, e, f, g,
h), (b, c, d, e, f, g, h)$\}$.

3) Find the number of all possible outcomes for the following experiments:

a) rolling three dice; generalization: rolling n dice;

b) spinning a slot machine with four reels having eight symbols each; generalization: spinning a slot machine with n reels of m symbols each;

c) dealing a player three cards from a 52-card deck;

d) dealing two players two cards each from 50 cards;

e) a race with nine competitors.

Solution:

a) Rolling a die has six possible outcomes, so rolling the three dice has $6 \times 6 \times 6 = 216$ possible outcomes.

b) Spinning a reel has eight possible outcomes, so spinning the four reels has $8 \times 8 \times 8 \times 8 = 4096$ possible outcomes.

Generalization: we have $\underbrace{m \times m \times ... \times m}_{n \text{ times}} = m^n$ possible outcomes.

c) The result is given by the number of 3-size combinations from 52, namely, $C_{52}^3 = \dfrac{50 \cdot 51 \cdot 52}{2 \cdot 3} = 22100$.

d) The two players are dealt a double card combination (xy)(zt) from the 50 cards. The result is given by the number of these combinations, which is $C_{50}^2 C_{48}^2 = \dfrac{49 \cdot 50}{2} \cdot \dfrac{47 \cdot 48}{2} = 1381800$.

e) The result is given by the number of permutations of nine elements, namely, $9! = 2 \cdot 3 \cdot 4 \cdot 5 \cdot 6 \cdot 7 \cdot 8 \cdot 9 = 362880$.

4) Write the field of events attached to the following experiments:

a) tossing a coin;

b) drawing a ball from an urn with three balls;

c) drawing two balls from an urn with three balls.

Solution:

a) The sample space is $\Omega = \{H, T\}$ (H – heads, T – tails), so the field of events is $\Sigma = \mathcal{P}(\Omega) = \{\phi, \{H\}, \{T\}, \{H, T\}\}$.

b) Denoting the three balls by a, b, c, we have $\Omega = \{a, b, c\}$ and $\Sigma = \mathcal{P}(\Omega) = \{\phi, \{a\}, \{b\}, \{c\}, \{a, b\}, \{a, c\}, \{b, c\}, \{a, b, c\}\}$.

c) With the same denotations as from point b), we have:
$\Omega' = \{(a, b), (a, c), (b, c)\}$ and $\Sigma' = \mathcal{P}(\Omega') =$
$\{\phi, \{(a, b)\}, \{(a, c)\}, \{(b, c)\}, \{(a, b), (a, c)\}, \{(a, b), (b, c)\}, \{(a, c), (b, c)\}, \Omega'\}$

5) An urn contains thirty balls numbered 1 to 30. Relating to the experiment of drawing a ball, what can you state about the following events (elementary, compound, relations between them):

A – the number of the drawn ball is even;
B – the number of the drawn ball is a multiple of 4;
C – the number of the drawn ball is 5;
D – the number of the drawn ball is a multiple of 5;
E – the number of the drawn ball is a power of 5.

Solution:
The only elementary event is C; the other events are compound.
The following couples of events are incompatible: A and C, B and C.
We have the following inclusions: $B \subset A$, $E \subset D$, $C \subset D$, $C \subset E$.

6) Two cards are drawn from a 52-card deck. Consider the events:

A – two aces are drawn;
B – two cards higher than Q are drawn;
C – 5♦ and a J are drawn.

Decompose these events in elementary events and specify their numbers.

Solution:
The elementary events attached to this experiment are the occurrences of the 2-size combinations (couples) from the 52 cards.

$A = \{(AA)\} = \{(A♠A♣), (A♠A♣), (A♠A♦), (A♣A♥), (A♣A♦),$
$(A♥A♥)\}$. The set A has $C_4^2 = 6$ elements (elementary events).

$B = \{(xy),$ with x, y being K or $A\} = \{(A♠A♣), (A♠A♣), (A♠A♦),$
$(A♣A♥), (A♣A♦), (A♥A♥), (K♠K♣), (K♠K♣), (K♠K♦), (K♣K♥),$
$(K♣K♦), (K♥K♥), (K♠A♠), (K♠A♣), (K♠A♥), (K♠A♦), (K♣A♠),$
$(K♣A♣), (K♣A♥), (K♣A♦), (K♥A♠), (K♥A♣), (K♥A♥), (K♥A♦),$
$(K♦A♠), (K♦A♣), (K♦A♥), (K♦A♦)\}$.

The set B is the union of the mutually exclusive sets $\{(KK)\}$, $\{(AA)\}$ and $\{(AK)\}$ and has $C_4^2 + C_4^2 + 4 \cdot 4 = 28$ elements.

$C = \{(5\spadesuit\ J)\} = \{(5\spadesuit\ J\spadesuit), (5\spadesuit\ J\clubsuit), (5\spadesuit\ J\heartsuit), (5\spadesuit\ J\diamondsuit)\}$.
C has four elements.

7) Find the probability of getting a multiple of 2 at a die roll.

Solution:
The number of outcomes that are favorable to respective event is three (these are: $\{2\}$, $\{4\}$, $\{6\}$).

The number of equally possible outcomes is six, so the probability is $3/6 = 1/2 = 50\%$.

8) An urn contains five white balls and three black balls. Find the probability of the following events:
a) A – drawing a white ball;
b) B – drawing a black ball.

Solution:
a) The number of cases that are favorable to event A is five (drawing any of the five white balls) and the number of equally possible cases is eight. Therefore $P(A) = 5/8$.
b) Similarly, $P(B) = 3/8$.

9) There are three pairs of socks of different colors in a basket. Two socks are randomly extracted from the basket.
What is the probability of getting two socks of same color?

Solution:
The number of equally possible cases is the number of all 2-size combinations of socks, namely, $C_6^2 = 15$.

The number of favorable cases is three, because we have three pairs of socks having the same color.

Thus, the probability is $3/15 = 1/5 = 20\%$.

10) An urn contains balls on which are inscribed all 4-digit numbers that can be formed by using the figures 1, 2, 3, 4, each used once (one number on each ball). What is the probability of drawing a ball with a number having the figure 1 as the first digit?

Solution:
We have 4! balls (this is the number of numbers that can be formed by using four figures, each once). The number of equally possible cases is then 4!.
The numbers having the figure 1 as first digit have the form 1xyz, with x, y, z distinct and taking the values 2, 3, 4. Their number is 3!.
The number of favorable cases is then 3!.
The probability of the event to be measured is
3!/4! = 1 x 2 x 3 / (1 x 2 x 3 x 4) = 1/4 = 25%.

11) An urn contains four white balls and six black balls. Two balls are drawn simultaneously. Find the probability of the events:
a) A – drawing two white balls;
b) B – drawing two black balls;
c) C – drawing two balls of the same color.

Solution:
The number of possible cases is C_{10}^2.

a) The number of cases that are favorable to event A is C_4^2;
therefore, $P(A) = \dfrac{C_4^2}{C_{10}^2} = 6/45 = 2/15$.

b) Similarly, $P(B) = \dfrac{C_6^2}{C_{10}^2} = 15/45 = 1/3$.

c) We have $C = A \cup B$. The events A and B are incompatible, so
$P(C) = P(A \cup B) = P(A) + P(B) = \dfrac{6+15}{45} = 21/45 = 7/15$.

12) Two dice, one red and one blue, are rolled. Consider the events:
A – occurrence of a number less than 4 on the red die;
B – occurrence of a number less than 3 on the blue die.
Find $P(A \cup B)$.

Solution:

The cases that are favorable to *A* are {1}, {2} and {3}; therefore, P(A) = 3/6.

The cases that are favorable to *B* are {1} and {2}; therefore, P(A) = 2/6.

The cases that are favorable to $A \cap B$ correspond to the ordered pairs (1, 1), (1, 2), (2, 1), (2, 2), (3, 1), (3, 2), and total six, in a probability field where the number of equally possible cases is 6 x 6 = 36.

We then have $P(A \cap B) = 6/36$.

The requested probability is

$$P(A \cup B) = P(A) + P(B) - P(A \cap B) = \frac{3}{6} + \frac{2}{6} - \frac{6}{36} = 24/36 = 2/3.$$

13) Five cards are drawn at once from a 32-card deck. What is the probability of the five cards containing at least one queen (*Q*)?

Solution:

Denoting by *A* the event to be measured *the five extracted cards contain at least one Q*, we then calculate the probability of the contrary event A^C – *the five extracted cards contain no Q*.

The equally possible elementary events are the occurrences of 5-size combinations of cards from the 32, a total of C_{32}^5.

The combinations that are favorable to event A^C have the form (xyztv), with *x*, *y*, *z*, *t*, *v* taking any card as value, except the four *Q*-cards. They total $C_{32-4}^5 = C_{28}^5$.

We then have:

$$P(A) = 1 - P(A^C) = 1 - \frac{C_{28}^4}{C_{32}^5} = 1 - \frac{25 \cdot 13 \cdot 9 \cdot 7}{28 \cdot 29 \cdot 31 \cdot 8} = 0.89832.$$

The problem could also be solved in this way:

Consider the events A_i – *the five extracted cards contain exactly* i *Q-cards*, i = 1, 2, 3, 4.

These events are mutually exclusive. We must calculate the probability of each of them (by counting the favorable combinations) and then add them together.

14) We have two urns, the first containing three white balls and four black balls and the second three white balls and five black balls.
A ball is drawn from a randomly chosen urn.
Find the probability for the drawn ball to be white.

Solution:
Denote the events:
A – the first urn is the chosen one;
B – the second urn is the chosen one;
C – the drawn ball is white.
A and B form a complete system of events and $P(A) = P(B) = 1/2$.
We have $P_A(C) = 3/7$ and $P_B(C) = 3/8$.
According to total probability formula, we have:

$$P(C) = P(A)P_A(C) + P(B)P_B(C) = \frac{1}{2} \cdot \frac{3}{7} + \frac{1}{2} \cdot \frac{3}{8} = 45/112 = 0.40178.$$

15) There are three urns. The first contains four green balls and two yellow balls; the second contains three green balls and three yellow balls; and the third contains five green balls and one yellow ball. An urn is randomly chosen and a ball is drawn.
Find the probability of drawing a green ball.

Solution:
Denote the following events:
A_i – urn number i is chosen, $i = 1, 2, 3$;
A – the drawn ball is green.
The events A_i form a complete system of events.
We have: $P(A_1) = P(A_2) = P(A_3) = 1/3$, $P_{A_1}(A) = 4/6$,
$P_{A_2}(A) = 3/6$, $P_{A_3}(A) = 5/6$.
By applying total probability formula, we find that:

$$P(A) = P(A_1)P_{A_1}(A) + P(A_2)P_{A_2}(A) + P(A_3)P_{A_3}(A) =$$

$$= \frac{1}{3} \cdot \frac{4}{6} + \frac{1}{3} \cdot \frac{3}{6} + \frac{1}{3} \cdot \frac{5}{6} = 2/3.$$

16) In a batch of 100 pieces, six pieces have remediable fabrication faults and four are rejects. Ten pieces are randomly selected from this batch. What is the probability of seven of the ten pieces being good, two having a remediable fault and one being a reject?

Solution:
By applying the scheme of the nonreturned ball for n = 10 and

s = 3, we find $P(10; 7, 2, 1) = \dfrac{C_{90}^7 C_6^2 C_4^1}{C_{100}^{10}} = 0.02589$.

17) An urn contains five white balls and five black balls. Two balls are randomly drawn in succession, without either ball being returned to the urn.
  What is the probability of obtaining the various results?

Solution:
We apply the scheme of the nonreturned ball for n = 2 and s = 2.
For drawing two white balls, the probability is
$$P(2; 2, 0) = \frac{C_5^2 C_5^0}{C_{10}^2} = 2/9 .$$
For drawing one white ball and one black ball, the probability is
$$P(2; 1, 1) = \frac{C_5^1 C_5^1}{C_{10}^2} = 5/9 .$$
For drawing two black balls, the probability is
$$P(2; 0, 2) = \frac{C_5^0 C_5^2}{C_{10}^2} = 2/9 .$$

18) Three shooters fire on a target. The first hits the target with a probability of 3/4, the second with a probability of 4/5 and the third with a probability of 5/6.
  a) What is the probability for the target to be hit three times?
  b) What is the probability for the target to be hit exactly two times?
  c) What is the probability for the target to be hit at least once?

Solution:

We are in the case of a Poisson scheme here, with $n = 3$,

$p_1 = 3/4$, $q_1 = 1/4$, $p_2 = 4/5$, $q_2 = 1/5$, $p_3 = 5/6$, $q_3 = 1/6$.

$$\varphi_3(z) = \left(\frac{3}{4}z + \frac{1}{4}\right)\left(\frac{4}{5}z + \frac{1}{5}\right)\left(\frac{5}{6}z + \frac{1}{6}\right).$$

a) The probability of the event to be measured is the coefficient of z^3 from the development of $\varphi_3(z)$, namely,

$$P_{3,3} = \frac{3}{4} \cdot \frac{4}{5} \cdot \frac{5}{6} = 1/2 = 50\%.$$

b) The probability of the event to be measured is the coefficient of z^2 from the development of $\varphi_3(z)$, namely,

$$P_{2,3} = \frac{3}{4} \cdot \frac{4}{5} \cdot \frac{1}{6} + \frac{3}{4} \cdot \frac{5}{6} \cdot \frac{1}{5} + \frac{4}{5} \cdot \frac{5}{6} \cdot \frac{1}{4} = 47/120 = 39.166\%.$$

c) The probability of the contrary event (the target not being hit) is the free term from the development of $\varphi_3(z)$, namely,

$$P_{0,3} = \frac{1}{4} \cdot \frac{1}{5} \cdot \frac{1}{6} = 1/120.$$

Thus, the searched probability is

$1 - P_{0,3} = 1 - 1/120 = 119/120 = 99.166\%.$

19) A die is rolled eight times. What is the probability of rolling a 5 exactly five times?

Solution:

We are in case of a Bernoulli scheme here, with $n = 8$ and $m = 5$.
We have $p = 1/6$, $q = 5/6$ and then

$$P_{5,8} = C_8^5 \left(\frac{1}{6}\right)^5 \left(\frac{5}{6}\right)^3 = \frac{56 \cdot 125}{6^8} = 0.00416.$$

20) A pair of dice is rolled seven times. What is the probability of getting a total of five points exactly two times?

Solution:

The probability of getting a total of five points at one roll of the two dice is $4/36 = 1/9$ because we have four favorable cases (that is, the ordered pairs $(1, 4)$, $(4, 1)$, $(2, 3)$, $(3, 2)$) from $6 \times 6 = 36$ equally possible cases.

We are in case of a Bernoulli scheme here, with $n = 7$ and $m = 2$. We have $p = 1/9$, $q = 8/9$ and then

$$P_{2,7} = C_7^2 \left(\frac{1}{9}\right)^2 \left(\frac{8}{9}\right)^5 = \frac{21 \cdot 8^5}{9^7} = 0.14387.$$

21) An urn contains eleven balls: three white balls, five green balls and three red balls. Seven draws of one ball each are performed, and each drawn ball is put back in the urn. Find the probability of getting two white balls, four green balls and one red ball in all seven draws.

Solution:

We are in case of a polynomial scheme here, with $n = 7$ and $s = 3$. We have $p_1 = 3/11$, $p_2 = 5/11$, $p_3 = 3/11$ and the searched probability is

$$P(7; 2, 4, 1) = \frac{7!}{2! \cdot 4! \cdot 1!} \left(\frac{3}{11}\right)^2 \left(\frac{5}{11}\right)^4 \left(\frac{3}{11}\right)^1 = \frac{105 \cdot 3^3 \cdot 5^4}{11^7} = 0.09092.$$

22) Three dice are rolled. Calculate the probability of getting at least one double (at least two identical numbers showing on the three dice).

Solution:

To calculate the probability for a specific number to come up on at least two dice, let us consider number 1 and the events:

B_1 – number 1 comes up on the first and on the second die,

B_2 – number 1 comes up on the second and on the third die

B_3 – number 1 comes up on the first and on the third die.

We have $P(B_i) = 1/36$ $(i = 1, 2, 3)$.

We have three intersections of two sets B_i. Each intersection has a single element, namely, the variant $(1, 1, 1)$, which has the probability $1/(6 \times 6 \times 6) = 1/216$, so
$$P(B_1 \cap B_2) = P(B_2 \cap B_3) = P(B_1 \cap B_3) = 1/216.$$

The intersection $B_1 \cap B_2 \cap B_3$ still has a single element and
$$P(B_1 \cap B_2 \cap B_3) = 1/216.$$

According to the inclusion-exclusion principle, we have
$$P(B_1 \cup B_2 \cup B_3) = 3 \cdot 1/36 + 3 \cdot 1/216 + 1/216 = 2/271/216.$$

The same ratiocination is valid for any other number chosen.

The events *number* i *comes up on at least two dice* $(i = 1, \ldots, 6)$ are mutually exclusive, so the probability of getting any double is
$$P = 6 \cdot 2/27 = 4/9.$$

23) At a slot machine having four reels and seven symbols, calculate the probability of getting a double (exactly two identical symbols on the four reels).

Solution:
Let us calculate first the probability of getting a specific double. Let s be a symbol and consider the events:

$C_1 - s$ only comes up on reels 1 and 2 – the variants (ssxy);

$C_2 - s$ only comes up on reels 1 and 3 – the variants (sxsy);

$C_3 - s$ only comes up on reels 1 and 4 – the variants (sxys);

$C_4 - s$ only comes up on reels 2 and 3 – the variants (xssy);

$C_5 - s$ only comes up on reels 2 and 4 – the variants (xsys);

$C_6 - s$ only comes up on reels 3 and 4 – the variants (xyss), with $x, y \neq s$.

$$P(C_i) = \frac{6 \cdot 6}{7 \cdot 7 \cdot 7 \cdot 7}, \text{ for every } i = 1, \ldots, 6.$$

The intersections of sets C_i are empty; therefore,

$$P(C_1 \cup \ldots \cup C_6) = 6 \cdot \frac{6 \cdot 6}{7 \cdot 7 \cdot 7 \cdot 7} = \frac{6^3}{7^4}.$$

Let us see what the probability is of getting any double.
Denoting the symbols by $s_1, \ldots, s_7$ and the events

D_i – *getting a double* s_i, i = 1, ..., 7, we observe that the intersections of 3, 4, ..., and 7 sets D_i are empty (if three sets D_i have a common element, the respective variant should contain 3 x 2 = 6 symbols, which is impossible).

We have C_7^2 intersections of two sets D_i and each has six elements (denoting symbolically by 1 and 2 two different symbols, the variants belonging to the intersection are (1122), (1221), (1212), (2112), (2121) and (2211), a total of 6).

We then have $P(D_i \cap D_j) = \dfrac{6}{7^4}$, for any i, j = 1, ..., 21, i ≠ j

According to the inclusion-exclusion principle, we have:

$$P(D_1 \cup ... \cup D_7) = 7 \cdot \frac{6^3}{7^4} - 21 \cdot \frac{6}{7^4} = 0.57725 = 57.725\%.$$

24) At a blackjack game, calculate the probability for a player to get a total of twenty points from the first two cards (provided no other cards are shown), in two cases: a) a 52-card deck is used; b) two 52-card decks are used.

Solution:
a) for one deck

The variants totaling twenty points are of the type A + 9 or 10 + 10 (as a value; that is, any 2-size combination of cards from 10, J, Q, K).

We have sixteen variants A + 9 (4 aces and 4 nines) and $C_{16}^2 = 120$ variants 10 + 10 (all 2-size combinations of cards from the sixteen cards with a value of 10).

The number of all possible distribution variants for two cards is $C_{52}^2 = 1326$. The probability is then $P = \dfrac{16+120}{1326} = 68/663$.

b) for two decks

The variants of type A + 9 number sixty-four and those of type 10 + 10 (as value) number $C_{32}^2 = 496$.

The number of all possible distribution variants is $C_{104}^2 = 5356$.

The probability is then $P = \dfrac{64+496}{5356} = 140/1339$.

25) You are participating in a Texas Hold'em Poker game with seven opponents and you are dealt $K3$. Calculate the probability of your opponents holding no K (before the flop).

Solution:
We calculate the probability of the contrary event
A – *at least one opponent holds at least one K.*
We first calculate the probability of a fixed opponent holding at least one K.

The combinations that are favorable for this event are (Kx), with x different from the cards in your hand. To count them, we split them into two groups:

(Kx), x different from K – and number $3 \times (52 - 2 - 1 - 2) = 3 \times 47 = 141$, and

(KK) – and number $C_3^2 = 3$.

In total, we have $141 + 3 = 144$ favorable combinations, from $C_{50}^2 = 1225$ possible, so the probability is $144/1225$.

Denoting by A_i the events *opponent number* i *holds at least one K* ($i = 1, ..., 7$), we have $P(A_i) = 144/1225$ and can then study the intersections of 2, 3, 4,5, 6 and 7 sets A_i:

There are three K-cards still in play, so more than three opponents cannot hold at least one K. Thus, the intersections of more than three sets A_i are empty.

We have $C_7^2 = 21$ intersections for two sets A_i.

Such intersection contains double combinations of the type (Kx)(Ky), with x and y different from the cards in your hand.

We split them into three groups:

(Kx)(Ky), with x and y different from K – and number $3 \times (52 - 2 - 1 - 2) \times 2 \times (52 - 2 - 2 - 1 - 1) = 12972$;

(KK)(Ky), with y different from K – and number $C_3^2 \times 1 \times (52 - 2 - 2) = 144$, and

(Kx)(KK), with x different from K – and number $3 \times (52 - 2 - 1 - 1) \times C_2^2 = 144$.

In total, we have $12972 + 144 + 144 = 13260$ combinations that are favorable from $C_{50}^2 C_{48}^2$ possible, so $P(A_i \cap A_j) = \dfrac{13260}{C_{50}^2 C_{48}^2}$, for any $i, j = 1, ..., 21, i \neq j$.

We have $C_7^3 = 35$ intersections of three sets A_i.

Such intersection contains triple combinations of the type $(Kx)(Ky)(Kz)$, with x, y and z different from K, and number $3 \times (52 - 2 - 1 - 2) \times 2 \times (52 - 2 - 2 - 1 - 1) \times 1 \times$ $\times (52 - 2 - 2 - 2 - 1) = 583740$.

The number of all possible triple combinations is $C_{50}^2 C_{48}^2 C_{46}^2$, so we have $P(A_i \cap A_j \cap A_k) = \dfrac{583740}{C_{50}^2 C_{48}^2 C_{46}^2}$, for any $i < j < k$, $i, j, k = 1, ..., 35$.

We can now apply the inclusion-exclusion principle:

$$P(A) = 7 \cdot \frac{144}{1225} - 21 \cdot \frac{13260}{C_{50}^2 C_{48}^2} + 35 \cdot \frac{583740}{C_{50}^2 C_{48}^2 C_{46}^2} = 0.63562.$$

The probability of the contrary event (requested by the problem) is $1 - P(A) = 0.36438$.

26) You are participating in a Texas Hold'em Poker game and the flop cards are dealt.

Find the probability for you to achieve a four-of-a-kind formation by the river from among the cards in your hand and the flop cards.

Solution:

A specific four-of-a-kind formation can be denoted by (CCCCx), where C is the four-of-a-kind card.

Let us denote by A the event: *You achieve a (CCCCx) four-of-a-kind formation, after all community cards are dealt by using the five cards in your hand and the community cards, and the number of cards from each does not matter.*

A is equal to *the last two community cards to come allow you to achieve a four-of-a-kind formation.*

We calculate the probability $P(A)$, at the point in the game right after the flop cards are dealt.

The only variable the probability depends on is c = number of seen C-cards (from your own hand and the community cards).

This variable must fit the following initial conditions: c is a natural number and $0 \le c \le 4$.

For a (CCCCx) four-of-a-kind formation to be finally achieved, a minimal initial condition is that the cards showing must contain a minimum of two C cards.

We have five cases to study here:

1) $c = 0$. Then $P(A) = 0$ (initial condition not respected).

2) $c = 1$. Then $P(A) = 0$ (for the same reason as in case 1)).

3) $c = 2$. The cards showing are (CCxyz), with x, y, z different from C.

The last two cards to come (the turn and river) are 2-size combinations of cards from the forty-seven unseen cards.

The favorable combinations for A to occur are (CC), and number $C_{4-c}^2 = (3-c)(4-c)/2$.

The number of all possible combinations for turn and river cards is $C_{47}^2 = 1081$.

Therefore, the probability is
$P(A) = (3-c)(4-c)/2162 = 1/1081$.

4) $c = 3$. The cards showing are (CCCxy), with x, y different from C.

The favorable combinations are (Cx), and number
$(4-c)[47-(4-c)] = (4-c)(43+c)$.

The probability is $P(A) = (4-c)(43+c)/1081 = 46/1081$.

5) $c = 4$. The cards showing are (CCCCx). Any combination is favorable (the four-of-a-kind formation is already achieved); therefore, the probability is $P(A)=1$.

27) In a Texas Hold'em Poker game, calculate the probability of an unpaired hand (two different cards as value) improving to a pair on the flop.

Solution:

We calculate the probability of improving to exactly a pair (no trips, two pairs or quads).

Let us denote by (CD) the pocket cards (*C* different from *D* as value).

The favorable combinations of the flop are:

(Cxy), *x*, *y* different from *C* and *D*, *x* different from *y* – and number $3(C^2_{50-3-3} - 11C^2_4)$;

(Dxy), *x*, *y* different from *C* and *D*, *x* different from *y* – and number $3(C^2_{50-3-3} - 11C^2_4)$.

In total, we have 5280 favorable combinations from $C^3_{50} = 19600$ possible.

The probability of the event to be measured is 5280/19600 = 26.938%.

28) In a Texas Hold'em Poker game, calculate the probability of the flop being unpaired.

Solution:

– From the outside view (no pocket cards are showing):

The favorable combinations are (ABC), with *A*, *B* and *C* mutually different (as value), and number $C^3_{13} \times 4 \times 4 \times 4$ (three each of the thirteen values, four cards each), from C^3_{52} possible.

The probability is then

$$C^3_{13} \times 4 \times 4 \times 4 / C^3_{52} = 18304 / 22100 = 82.823\% .$$

– From the inside view (you are participating in the game), we must take into account your pocket cards:

If you hold a pair—let it be (EE)—the favorable combinations of the flop are (BCD), with *B*, *C* and *D* mutually different (*B*, *C*, or *D* could take the value *E*). These can be split as follows:

(BCD), with *B*, *C*, *D* different from *E*, numbering $C^3_{13} \times 4 \times 4 \times 4$, and

(EBC), with *B*, *C* different from *E*, numbering $2 \times C^2_{12} \times 4 \times 4$.

In total, we have 9008 favorable combinations from 22100 possible.

The probability is then 9008/22100 = 40.760%.

If you hold two unpaired cards—let them be (AB)—the favorable combinations of the flop are (CDE), with C, D and E mutually different. These can be split as follows:

(CDE), with C, D, E different from A and B, numbering $C_{13}^3 \times 4 \times 4 \times 4$;

(ACD), with C, D different from A and B, numbering $3 \times C_{11}^2 \times 4 \times 4$;

(BCD), with C, D different from A and B, numbering $3 \times C_{11}^2 \times 4 \times 4$;

(ABC), with C different from A and B, numbering $3 \times 3 \times 11 \times 4$.

In total, we have 16104 favorable combinations from 22100 possible.

The probability is then $16104/22100 = 72.868\%$.

29) You are participating in a Texas Hold'em Poker game with n opponents, you are dealt two suited cards (cards with the same symbol) and the flop comes with three additional cards of your suit.

Find the general formula of probability of none of your opponents holding two cards of your suit.

Solution:
We have $52 - 2 - 3 = 47$ unseen cards.

There are still $13 - 2 - 3 = 8$ cards of your suit among the unseen cards.

Let us denote by S the symbol of your suit.

We find first the probability of a specific opponent holding (SS) (denote this event by A):

The total number of possible 2-size combinations that an opponent can be dealt is $C_{47}^2 = 1081$.

The number of favorable combinations (SS) is $C_8^2 = 28$.

The probability is then $P(A) = C_8^2 / C_{47}^2 = 28/1081$.

Denote by A_i the events *opponent number* i *holds (SS)*,

$i = 1, ..., n$. The event to be measured is $\bigcup_{i=1}^{n} A_i$.

We have $C_n^2 = n(n-1)/2$ intersections of two events A_i.

Such an intersection contains the double combinations (SS)(SS), and number $C_8^2 C_6^2$. The number of all double combinations two opponents can be dealt is $C_{47}^2 C_{45}^2$.

Thus, $P\left(A_i \cap A_j\right) = \dfrac{C_8^2 C_6^2}{C_{47}^2 C_{45}^2}$, for any $i \neq j$, $i, j = 1, ..., n$.

We have $C_n^3 = n(n-1)(n-2)/6$ intersections for each of three events A_i. Such an intersection contains the triple combinations (SS)(SS)(SS), and numbers $C_8^2 C_6^2 C_4^2$.

The number of all triple combinations three opponents can be dealt is $C_{47}^2 C_{45}^2 C_{43}^2$.

Thus, $P\left(A_i \cap A_j \cap A_k\right) = \dfrac{C_8^2 C_6^2 C_4^2}{C_{47}^2 C_{45}^2 C_{43}^2}$, for any $i < j < k$,

$i, j, k = 1, ..., n$.

We have $C_n^4 = n(n-1)(n-2)(n-3)/24$ intersections of four events A_i.

Such an intersection contains the quadruple combinations (SS)(SS)(SS)(SS), and numbers $C_8^2 C_6^2 C_4^2 C_2^2 = C_8^2 C_6^2 C_4^2$.

The number of all quadruple combinations four opponents can be dealt is $C_{47}^2 C_{45}^2 C_{43}^2 C_{41}^2$.

Thus, $P\left(A_i \cap A_j \cap A_k \cap A_h\right) = \dfrac{C_8^2 C_6^2 C_4^2}{C_{47}^2 C_{45}^2 C_{43}^2 C_{41}^2}$, for any

$i < j < k < h$, $i, j, k, h = 1, ..., n$.

A maximum of four opponents can simultaneously hold (SS), because there are only 8 S-cards in play; therefore, the intersections of more than four events A_i are empty.

We can now apply the inclusion-exclusion principle:

$$P\left(\bigcup_{i=1}^n A_i\right) = \frac{n C_8^2}{C_{47}^2} - \frac{C_n^2 C_8^2 C_6^2}{C_{47}^2 C_{45}^2} + \frac{C_n^3 C_8^2 C_6^2 C_4^2}{C_{47}^2 C_{45}^2 C_{43}^2} - \frac{C_n^4 C_8^2 C_6^2 C_4^2}{C_{47}^2 C_{45}^2 C_{43}^2 C_{41}^2}$$ and this

is the searched formula.

Exercise: Do the complete combinatorial calculus and do the algebraic calculus to put the above expression in a polynomial form.

Unsolved applications

1) Determine $\mathcal{P}(\Omega)$ if:

a) $\Omega = \{0, 1, 2\}$;

b) $\Omega = \{A, B\}$;

c) $\Omega = \{\{a, b\}, \{c, d\}, \{a, b, c\}, \{e\}\}$.

2) Write the sample space of the following experiments:

a) draw of a ball from an urn containing eight balls;

b) draw of two balls from two urns (one ball from each), the first containing four yellow balls and the second three black balls;

c) draw of a card from a 32-card deck (from 7 upward);

d) rolling three dice;

e) choosing four numbers from among the numbers 5, 7, 9, 11, 13, 15;

f) choosing eight letters from among the letters $m, n, o, p, q, r, s, t, u, v$;

3) Find the number of all possible outcomes for the following experiments:

a) tossing five coins; generalization: tossing n coins;

b) spinning a slot machine with five reels and ten symbols;

c) dealing a player four cards from a 52-card deck;

d) dealing two players two cards each from 52 cards;

e) dealing three players three cards each from 52 cards;

f) a race with ten competitors.

4) Write the field of events attached to the following experiments:

a) choosing a card from a 24-card deck;

b) drawing a ball from an urn with five balls;

c) drawing five balls from an urn with seven balls.

5) An urn contains fifty balls numbered 1 to 50.

Relating to the experiment of drawing a ball, what can you state about the following events (elementary, compound, relations between them):

A – the number of the drawn ball is even;
B – the number of the drawn ball is a multiple of 4;
C – the number of the drawn ball is 5;
D – the number of the drawn ball is a multiple of 5;
E – the number of the drawn ball is a power of 5;
F – the number of the drawn ball is a multiple of 10;
G – the number of the drawn ball is a multiple of 3;
H – the number of the drawn ball is a power of 3;
I – the number of the drawn ball is even.

6) Two cards are drawn from a 52-card deck. Consider the events:

A – two clubs are drawn;
B – two cards having a value less than 5 are drawn;
C – a 7 and a Q are drawn.

Decompose these events in elementary events and specify their numbers.

7) Find the probability of getting a multiple of 3 at a die roll.

Find the probability of getting a total of 5 points when rolling two dice. Find the probability of getting a total of 10 points when rolling three dice.

8) An urn contains nine white balls and four black balls. Find the probability of the following events:

a) A – drawing a white ball;
b) B – drawing a black ball.

9) In a pencil box are five pairs of pencils of same length (five separate lengths). Two pencils are randomly drawn from the box.

What is the probability of drawing a pair of pencils of the same length?

10) In an urn are balls on which are inscribed all 5-digit numbers that can be formed by using the figures 1, 2, 3, 4, 5, 6, 7, each used once (one number on each ball). What is the probability of drawing a ball having a number with the figure 2 as first digit? How about the probability of drawing a ball having a number with the figure 5 as last digit?

11) An urn contains seven white balls and ten black balls. Three balls are drawn simultaneously. Find the probability of the events:
a) A – drawing exactly two white balls;
b) B – drawing exactly two black balls;
c) C – drawing three balls of the same color.

12) Two dice, one green and one white, are rolled. Consider the events:
A – occurrence of a number less than 5 on the green die;
B – occurrence of a number less than 4 on the white die.
Find $P(A \cup B)$.

13) Five cards are draw at once from a 52-card deck. What is the probability of the five cards containing:
a) at least one king (K);
b) at least three hearts $(\heartsuit)$;
c) exactly three cards of same value (triple);
d) two aces and two jacks $(AAJJ)$;
e) at least two cards of same value (at least a pair).

14) We have three urns, the first containing four white balls and five black balls; the second two white balls and seven black balls; and the third three white balls and four black balls. A ball is drawn from a randomly chosen urn. Find the probability for the drawn ball to be white.

15) We have four urns, the first containing five green balls and three yellow balls; the second containing four green balls and four yellow balls; the third containing six green balls and two yellow balls; and the fourth containing three green balls and five yellow balls. An urn is randomly chosen and a ball is drawn. Find the probability of drawing a yellow ball.

16) In a batch of 200 pieces, eight pieces have remediable fabrication faults and seven are rejects. Twenty pieces are randomly taken from this batch. What is the probability of fifteen of them being good, three having remediable faults and two being rejects?

17) An urn contains seven white balls and eleven black balls.
Two balls are randomly drawn in succession, without either ball being returned to the urn. What is the probability of obtaining the various results?

18) Three shooters fire on a target. The first hits the target with a probability of 0.85, the second with a probability of 0.75 and the third with a probability of 0.90.
a) What is the probability for the target to be hit three times?
b) What is the probability for the target to be hit exactly two times?
c) What is the probability for the target to be hit at least once?

19) A die is rolled twelve times. What is the probability of rolling a 6 exactly three times? What is the probability of rolling an uneven number exactly five times?

20) A pair of dice is rolled eight times. What is the probability of getting a total of seven points exactly three times? What is the probability of getting two numbers higher than 4 exactly four times?

21) An urn contains twenty balls: five white balls, eight green balls and seven red balls. Eleven draws of one ball each are performed, and each time the ball is put back in the urn. Find the probability of getting three white balls, five green balls and three red balls in all eleven draws.

22) Four dice are rolled. Calculate the probability of getting at least one triple (at least three identical numbers on the four dice) and the probability of getting two doubles (two pairs of different numbers).

23) At a slot machine with five reels, each having nine symbols, calculate the probability of getting a triple (exactly three identical symbols on all five reels).

24) In a blackjack game, calculate the probability for a player to get a total of 19 points from the first two cards (provided no other cards are shown), in two cases: a) a 52-card deck is used; b) two 52-card decks are used.
Calculate the probability for the player to get a total of 20 points from the first three cards in both cases a) and b).

25) You are participating in a Texas Hold'em Poker game with nine opponents and you are dealt $J4$. Calculate the probability of your opponents holding no J and the probability of your opponents holding no J or 4 (before the flop).

26) You are participating in a Texas Hold'em Poker game and the flop cards are dealt. Find the probability for you to achieve a flush formation (five cards of same symbol) by the river, from your own cards and from the flop cards.

27) In a Texas Hold'em Poker game, calculate the probability of an unpaired hand (two different cards as value) improving to a triple on the flop.

28) In a Texas Hold'em Poker game, calculate the probability of the flop containing no suited cards (there are no two cards with the same symbol among the flop cards).

29) You are participating in a Texas Hold'em Poker game with n opponents, you are dealt two suited cards (cards with the same symbol) and the flop comes with exactly two cards of your suit.
Find the general formula for the probability of none of your opponents holding two cards of your suite (calculated in the moment right after the flop).

30) The twelve issues of a monthly magazine are randomly archived on a shelf. What is the probability for the magazines to be archived in chronological order of publication?

31) Each of coefficients of the equation $ax + b = 0$, $a \neq 0$ is obtained by rolling a die and registering the outcome.
Find the probability for the equation obtained to include an integer number as a solution.

32) At a competition are participating three competitors from town A, four competitors from town B and five competitors from town C.
What is the probability of the first three ranked being from the same town? What is the probability of all competitors from same town being the first ranked (in a compact group)?

33) Six children are seated in aleatory order on a bench.
What is the probability for two specific children to be seated next to each other?

34) A team of eleven students is randomly chosen from a class of twenty students.
What is the probability for a specific student to be chosen for the team?

35) At an exam, the probability for a candidate to pass the math is 0.9, physics is 0.8 and both subjects is 0.7. Find the probability of:
a) the candidate passing the math or the physics;
b) the candidate passing the math, if he or she passes the physics;
c) the candidate passing both subjects.

36) The probability for a light bulb to blow out once a month is 0.3. What is the probability of none of the fifteen light bulbs in a house blowing out in the next three months?

37) A secretary edits four letters and four addressed envelopes, and then randomly inserts all four letters into the envelopes, one in each envelope. Find the probability of:
a) each letter being inserted into the right envelope;
b) none of letters being inserted into the right envelope;
c) exactly two letters being inserted into the right envelopes.

38) Each of three persons holds a 52-card deck. Each of them randomly draws a card from his or her deck. Calculate the probability of:

  a) all three cards drawn being spades (♠);
  b) all three cards drawn having a black color (♠ or ♣);
  c) exactly two from the cards drawn being hearts (♥);
  d) at least one ace being drawn;
  e) drawing only cards with a value higher than 10.

APPLICATIONS AND RESULTS FOR THE GAMES OF CHANCE

This chapter contains a large collection of numerical probability results covering a very large part of the gaming events and situations encountered in games of chance.

The results are grouped into sections along with a description of each game and solutions to a few of the probability applications involved in each game.

The solved applications can be skipped by gamblers interested only in numerical results because the partial results are used again at the end of each subsection and listed, mostly in tables, from where they can be easily selected to match the desired situation.

On the other hand, readers who want to improve their knowledge of probability calculus may take this chapter as a new collection of problems and continue to study the solved applications.

But this chapter is primarily addressed to gamblers who want to use the numerical results to evaluate chances or make choices in specific gaming situations.

They may also find themselves discouraged in their addictive behavior when faced with the real winning odds that some games or gaming situations offer.

The numerical probability results are shown in a double format: fraction and decimal numbers as percentages; the second format always includes five decimal places for each irrational number.

The games of chance covered by this chapter are: Slots, Roulette, Blackjack and Texas Hold'em Poker, which are the most popular.

Gamblers interested in all games can also refer to the book titled *Probability Guide of Gambling* (which focuses on applying probability theory in gambling) for sections on these games as well as another five: Dice, Baccarat, Classical Poker, Lottery and Sports bets.

Gamblers using this collection of results must bear in mind that knowing the odds only results in additional information and not in immediate or even short-term winnings.

These results are just the numerical applications of mathematical probability in gambling.

Using them in a properly built strategy is not only a matter of mathematics but also of unavoidable subjective choice.

We recommend that all readers read the last chapter titled *The Probability-based Strategy*, as well as the section on *Relativity of probability* from first chapter, to see how to incorporate the role of probability theory in making decisions.

Slots

A slot machine is a device with three, four or five reels that is actuated mechanically or electrically after one or more coins or paid chips are inserted.

The reels have on their external side several graphical symbols, with the same symbols on each reel, and they spin independently at the same time, after the player activates the handle or the button.

After the reels stop spinning, a certain combination of collinear symbols occurs by the side of a central mark (the winning line).

Any winnings are differentiated by the combinations of symbols achieved.

Generally, a player wins when three or four identical symbols line up, or with other combinations that hold two identical symbols.

The slot machine is also called a *flipper, fruit machine* or *one-armed bandit.*

There are also electronic variants of this system that randomly generate symbols on a display.

We now present the probabilities of obtaining winning combinations for slot machines having three and four reels and one winning line, which are the most common.

Three reels

At a three-reel slot machine, the winning combinations are three identical symbols (triple), combinations with a certain symbol and two certain identical symbols and two identical symbols (double).

The problem of calculating the probabilities of occurrence of the various combinations is most similar to the one from the dice case.

Let n be the number of symbols on each reel.

The total number of possible combinations that can occur after the machine is actuated is n^3.

The probability of obtaining a certain triple is $(1/n)(1/n)(1/n) = 1/n^3$, and the probability of obtaining a triple is $n \cdot (1/n^3) = 1/n^2$ because the events *a triple of symbol* s_i *occurs,* $i = 1, ..., n$, are mutually exclusive.

Probability of obtaining a certain triple is $P = 1/n^3$.
Probability of obtaining a triple is $P = 1/n^2$.

Let us calculate the probability of obtaining a certain double. Let s be a symbol and let:

A_1 – s only appears on reels 1 and 2 – variants (ssx), $x \neq s$

A_2 – s only appears on reels 1 and 3 – variants (sxs), $x \neq s$

A_3 – s only appears on reels 2 and 3 – variants (xss), $x \neq s$

be the events involved.

$P(A_1) = P(A_2) = P(A_3) = (n-1)/n^3$.

The intersections of events A_i are empty; therefore,

$P(A_1 \cup A_2 \cup A_3) = 3(n-1)/n^3$.

The probability of obtaining a double is

$n \cdot 3(n-1)/n^3 = 3(n-1)/n^2$ because the events *a double of symbol* s_i *occurs,* $i = 1, \ldots, n$, are mutually exclusive (a variant containing two symbols s_i cannot contain other two symbols s_j, if $i \neq j$).

Probability of obtaining a certain double is $P = 3(n-1)/n^3$.

Probability of obtaining a double is $P = 3(n-1)/n^2$.

Let us now calculate the probability of obtaining a certain double (of symbol s) and a certain symbol (v).

Because there are only three favorable variants, namely, (ssv), (svs) and (vss), the probability is $3/n^3$.

Probability of obtaining a certain double and a certain symbol is $P = 3/n^3$.

Four reels

At a four-reel slot machine, the winning combinations are four identical symbols, triples and various combinations holding doubles.

The probability of obtaining four identical certain symbols is $(1/n)(1/n)(1/n)(1/n) = 1/n^4$, and the probability of obtaining a combination of four identical symbols is $n \cdot (1/n^4) = 1/n^3$.

Let us see what is the probability of obtaining a certain triple. Let s be a symbol and let:

$B_1 - s$ only appears on reels 1, 2 and 3 – variants (sssx)

$B_2 - s$ only appears on reels 1, 2 and 4 – variants (ssxs)

$B_3 - s$ only appears on reels 2, 3 and 4 – variants (xsss)

$B_4 - s$ only appears on reels 3, 4 and 1 – variants (sxss), $x \neq s$

be the events involved.

We have $P(B_i) = (n-1)/n^4$, $i = 1, 2, 3, 4$.

The intersections of events B_i are empty; therefore,

$P(B_1 \cup B_2 \cup B_3 \cup B_4) = 4(n-1)/n^4$.

Probability of obtaining a triple is $n \cdot 4(n-1)/n^4 = 4(n-1)/n^3$.

We can easily calculate the probability of obtaining a certain triple (of symbol s) and a certain symbol (v).

The favorable variants are four in number, that is, (sssv), (ssvs), (svss) and (vsss), so the probability is $4/n^4$.

Probability of obtaining 4 identical certain symbols is $P = 1/n^4$.

Probability of obtaining 4 identical symbols is $P = 1/n^3$.

Probability of obtaining a certain triple is $P = 4(n-1)/n^4$.

Probability of obtaining a triple is $P = 4(n-1)/n^3$.

Probability of obtaining a certain triple and a certain symbol is $P = 4/n^4$.

The probability of obtaining a certain double was calculated in detail in one of the solved applications from the previous chapter.

We have:

Probability of obtaining a certain double is $P = 6(n-1)^2/n^4$.

Probability of obtaining a double is $P = 3(n-1)(2n-5)/n^3$.

Probability of obtaining 2 certain doubles is $P = 6/n^4$.

Probability of obtaining 2 doubles is $P = 3(n-1)/n^3$.

That following tables list the probabilities of occurrence of the various winning combinations for slot machines having three and four reels, which have 5, 6, 7, 8, 9 or 10 symbols.

3 reels, n = 5

Event	P (/)	P (%)
certain triple	1/125	0.8
triple	1/25	4
certain double	12/125	9.6
double	12/25	48
certain double and certain symbol	3/125	2.4

3 reels, n = 6

Event	P (/)	P (%)
certain triple	1/216	0.46296
triple	1/36	2.77777
certain double	5/72	6.94444
double	5/12	41.66666
certain double and certain symbol	1/72	1.38888

3 reels, n = 7

Event	P (/)	P (%)
certain triple	1/343	0.29154
triple	1/49	2.04081
certain double	18/147	12.24489
double	18/49	36.73469
certain double and certain symbol	3/343	0.87463

3 reels, n = 8

Event	P (/)	P (%)
certain triple	1/512	0.19531
triple	1/64	1.5625
certain double	21/512	4.10156
double	21/64	32.8125
certain double and certain symbol	3/512	0.58593

3 reels, n = 9

Event	P (/)	P (%)
certain triple	1/729	0.13717
triple	1/81	1.23456
certain double	8/243	3.29218
double	8/27	29.62962
certain double and certain symbol	1/81	1.23456

3 reels, n = 10

Event	P (/)	P (%)
certain triple	1/1000	0.1
triple	1/100	1
certain double	27/1000	2.7
double	27/100	27
certain double and certain symbol	3/1000	0.3

4 reels, n = 5

Event	P (/)	P (%)
4 identical certain symbols	1/625	0.16
4 identical symbols	1/125	0.8
certain triple	16/625	2.56
triple	16/125	12.8
certain double	96/625	15.36
double	12/25	48
2 certain doubles	6/625	0.96
2 doubles	12/125	9.6

4 reels, n = 6

Event	P (/)	P (%)
4 identical certain symbols	1/1296	0.07116
4 identical symbols	1/216	0.46296
certain triple	5/324	1.54320
triple	5/54	9.25925
certain double	25/216	11.57407
double	35/72	48.61111
2 certain doubles	1/216	0.46296
2 doubles	5/72	1.38888

4 reels, n = 7

Event	P (/)	P (%)
4 identical certain symbols	1/2401	0.04164
4 identical symbols	1/343	0.29154
certain triple	24/2401	0.99958
triple	24/343	6.99708
certain double	216/2401	8.99625
double	162/343	47.23032
2 certain doubles	6/2401	0.24989
2 doubles	18/343	5.24781

4 reels, n = 8

Event	P (/)	P (%)
4 identical certain symbols	1/4096	0.02441
4 identical symbols	1/512	0.19531
certain triple	7/1024	0.68359
triple	7/128	5.46875
certain double	147/2048	7.17773
double	231/512	45.11718
2 certain doubles	3/2048	0.14648
2 doubles	21/512	4.10156

4 reels, n = 9

Event	P (/)	P (%)
4 identical certain symbols	1/6561	0.015241
4 identical symbols	1/729	0.13717
certain triple	32/6561	0.48773
triple	32/729	4.38957
certain double	128/2187	5.85276
double	104/243	42.79835
2 certain doubles	2/2187	0.09144
2 doubles	8/243	3.29218

4 reels, n = 10

Event	P (/)	P (%)
4 identical certain symbols	1/10000	0.01
4 identical symbols	1/1000	0.1
certain triple	9/2500	0.36
triple	9/250	3.6
certain double	243/5000	4.86
double	81/200	40.5
2 certain doubles	3/5000	0.06
2 doubles	27/1000	2.7

Roulette

Roulette is a casino game in which players place bets on numbers or groups of numbers; a player's goal is to predict the winning number or other of properties of that number (color, evenness, size, or place on the roulette table).

Each game consists of placing bets and waiting for a number, which is randomly generated by a spinning ball coming to rest inside a disk on which numbers are inscribed (the roulette wheel) that is also spinning, but in the opposite direction.

The bets are placed on the roulette table, which is designed to allow players to place paid chips on several combinations or sets of numbers.

There are 37 or 38 numbers in the game (European roulette has 37 numbers and American roulette has 38). The 38 numbers of the American roulette are 1, 2, 3, ..., 36, 0, 00.

Each of first 36 numbers (1–36) is marked with red or black and the color attached also corresponds to that number on the roulette wheel. The numbers 0 and 00 have no color.

The players bet by placing their chips in one or several fields on the roulette table.

When the winning number corresponds to a player's bet, he or she wins the amount bet multiplied by a coefficient approximately inversely proportional to the probability of winning that bet, with the approximation made in favor of the house.

For example, if a player places amount S in the **RED** cell and the winning number is 19 (red), that player wins amount S (1 – 1 winning) (the player receives S + S = 2S in return for winning).

If a player places amount S in the **First 12** cell and the winning number is 7, he or she wins amount 2S (2 – 1 winning) (the player receives S + 2S = 3S in return for winning).

If a player places amount S on the common side of the cells numbered 20 and 23 (two numbers are bet) and the winning number is 23, he or she wins amount 20S (20 – 1 winning) (the player receives S + 20S in return for winning).

Here are the simple bet possibilities, consisting of placing chips in a single field, along with that field and the winning level:

1 number – on the cell where a number is inscribed: **40 – 1**

2 numbers – on the common leg of two adjoining cells: **20 –1**

3 numbers – on the exterior leg of a three-number row: **13 – 1**

4 numbers – on the common corner of four adjoining cells: **9 – 1**

6 numbers – on the common corner of the exterior legs of two adjacent rows: **6 – 1**

12 numbers – on the bottom side of a column or cell corresponding to a 12-number rectangle: **2 – 1**

18 numbers – on a cell where a property is inscribed: red, black, even, odd, small or big: **1 – 1**.

The winning levels vary from one casino to another, but they are not much different from those described here.

The probabilities of winning the bets are approximately equal in both American and European roulette (the difference varies from $1/1406$ in the case of one number bet to about $1/78$ in the case of a color bet), so we present here the probabilities for a single case (American roulette).

The probabilities of the events attached to the experiment of spinning the roulette are easily calculable—each probability is given by the ratio between the number of favorable numbers and 38.

We present first the probabilities of winning simple bets, which consist of unique placements (in a single field of a roulette table):

1 certain number:	$P = 1/38 = 2.63157\%$
2 certain numbers:	$P = 1/19 = 5.26315\%$
3 certain numbers:	$P = 3/38 = 7.89473\%$
4 certain numbers:	$P = 2/19 = 10.52631\%$
6 certain numbers:	$P = 3/19 = 15.78947\%$
12 certain numbers:	$P = 6/19 = 31.57894\%$
18 certain numbers:	$P = 9/19 = 47.36842\%$

A player can also place complex bets, which consist of many simultaneous placements.

The most practiced complex bets are those corresponding to partially complementary events (red + few black numbers, odd + few even numbers, etc.).

These bets can increase the odds of winning, but they may also diminish the eventual winning amounts at the same time.

Example:

A player bets amount $2S$ on red color and amount S on number 6 (black).

A. In the case of a black number different from 6 winning, the player looses $3S$.

B. In the case of number 6 winning, the player wins $40S - 2S = 38S$.

C. In the case of a red number winning, the player wins $2S - S = S$.

D. In the case of 0 or 00 winning, the player looses $3S$.

The probability of winning is
$$P(B \cup C) = 1 - P(A \cup D) = 1 - 19/38 = 1/2 = 50\%.$$

If, along with the bet on red, several bets are placed on black numbers, the probability of winning will increase, but the eventual winning amount will decrease.

The optimum bet combinations are those having an acceptable ratio (arithmetically, but also according to a player's personal criteria) between the winning probabilities and the eventual winning and risked amounts.

We now present such partially complementary combinations, along with the probabilities of the events involved and the afferent winnings and losses.

1. **Bet amount *nS* on one color and amount *S* on each of two certain numbers of the opposite color.**

n is a variable coefficient to be chosen by the player.

Simply stated, we assume that the bet is *amount nS on red and amount S on each of two certain black numbers.*

The probability of a red number occurring is 18/38 = 9/19; in this case the amount won is $nS - 2S = (n - 2)S$.

The probability of one of the two black numbers occurring is 2/18 = 1/19; in this case the amount won is $40S - nS - S = (39 - n)S$.

The probability of 0, 00 or a black number different from those on which the bet was made occurring is $1 - 9/19 - 1/19 = 9/19$; in this case the amount lost is $nS + 2S = (n + 2)S$.

The probability of winning is 10/19 and the probability of losing is 9/19.

The first conclusion on coefficient n is that it must fit the condition $2 < n < 39$.

The next tables note the probabilities of the events attached to this experiment and the eventual afferent wins and losses for the coefficients $n = 3, 4, 5, 6, 7$.

n = 3

Event	P(/)	P(%)	Winning	Loss
winning the bet on color	9/19	47.36842	S	
winning a bet on a number	1/19	5.26315	36S	
not winning any bet	9/19	47.36842		5S

n = 4

Event	P(/)	P(%)	Winning	Loss
winning the bet on color	9/19	47.36842	2S	
winning a bet on a number	1/19	5.26315	35S	
not winning any bet	9/19	47.36842		6S

318

n = 5

Event	P(/)	P(%)	Winning	Loss
winning the bet on color	9/19	47.36842	3S	
winning a bet on a number	1/19	5.26315	34S	
not winning any bet	9/19	47.36842		7S

n = 6

Event	P(/)	P(%)	Winning	Loss
winning the bet on color	9/19	47.36842	4S	
winning a bet on a number	1/19	5.26315	33S	
not winning any bet	9/19	47.36842		8S

n = 7

Event	P(/)	P(%)	Winning	Loss
winning the bet on color	9/19	47.36842	5S	
winning a bet on a number	1/19	5.26315	32S	
not winning any bet	9/19	47.36842		9S

2. Bet amount nS on one color (red) and amount S on each of three certain numbers of the opposite color (black)

The probability of a red number occurring is 9/19; in this case the amount won is $nS - 3S = (n - 3)S$.

The probability of one of the three black numbers occurring is 3/38; in this case the amount won is $40S - nS - 2S = (38 - n)S$.

The probability of 0, 00 or a black number different from those on which the bet was made occurring is $1 - 9/19 - 3/38 = 17/38$; in this case the amount lost is $nS + 3S = (n + 3)S$.

The probability of winning is 21/38 and probability of losing is 17/38.

The condition on n is $3 < n < 38$.

The next tables note the probabilities of the events attached to this experiment and the eventual afferent wins and losses for the coefficients $n = 4, 5, 6, 7, 8$.

n = 4

Event	P(/)	P(%)	Winning	Loss
winning the bet on color	9/19	47.36842	S	
winning a bet on a number	3/38	7.89473	34S	
not winning any bet	17/38	44.73684		7S

n = 5

Event	P(/)	P(%)	Winning	Loss
winning the bet on color	9/19	47.36842	2S	
winning a bet on a number	3/38	7.89473	33S	
not winning any bet	17/38	44.73684		8S

n = 6

Event	P(/)	P(%)	Winning	Loss
winning the bet on color	9/19	47.36842	3S	
winning a bet on a number	3/38	7.89473	32S	
not winning any bet	17/38	44.73684		9S

n = 7

Event	P(/)	P(%)	Winning	Loss
winning the bet on color	9/19	47.36842	4S	
winning a bet on a number	3/38	7.89473	31S	
not winning any bet	17/38	44.73684		10S

n = 8

Event	P(/)	P(%)	Winning	Loss
winning the bet on color	9/19	47.36842	5S	
winning a bet on a number	3/38	7.89473	30S	
not winning any bet	17/38	44.73684		11S

3. Bet amount nS on one color (red) and amount S on each of four certain numbers of the opposite color (black)

The probability of a red number occurring is 9/19; in this case the amount won is $nS - 4S = (n - 4)S$.

The probability of one of the four black numbers occurring is 2/19; in this case the amount won is $40S - nS - 3S = (37 - n)S$.

The probability of 0, 00 or a black number different from those on which the bet was made occurring is $1 - 9/19 - 2/19 = 8/19$; in this case the amount lost is $nS + 4S = (n + 4)S$.

The probability of winning is 11/19 and probability of losing is 8/9.

The condition on n is $4 < n < 37$.

The next tables note the probabilities of the events attached to this experiment and the eventual afferent wins and losses for the coefficients $n = 5, 6, 7, 8, 9$.

$n = 5$

Event	P($/$)	P(%)	Winning	Loss
winning the bet on color	9/19	47.36842	S	
winning a bet on a number	2/19	10.52631	32S	
not winning any bet	8/19	42.10526		9S

$n = 6$

Event	P($/$)	P(%)	Winning	Loss
winning the bet on color	9/19	47.36842	2S	
winning a bet on a number	2/19	10.52631	31S	
not winning any bet	8/19	42.10526		10S

$n = 7$

Event	P($/$)	P(%)	Winning	Loss
winning the bet on color	9/19	47.36842	3S	
winning a bet on a number	2/19	10.52631	30S	
not winning any bet	8/19	42.10526		11S

n = 8

Event	P(/)	P(%)	Winning	Loss
winning the bet on color	9/19	47.36842	4S	
winning a bet on a number	2/19	10.52631	29S	
not winning any bet	8/19	42.10526		12S

n = 9

Event	P(/)	P(%)	Winning	Loss
winning the bet on color	9/19	47.36842	5S	
winning a bet on a number	2/19	10.52631	28S	
not winning any bet	8/19	42.10526		13S

4. Bet amount *nS* on one color (red) and amount *S* on each of five certain numbers of the opposite color (black)

The probability of a red number occurring is 9/19; in this case the amount won is $nS - 5S = (n - 5)S$.

The probability of one of the five black numbers occurring is 5/38; in this case the amount won is $40S - nS - 4S = (36 - n)S$.

The probability of 0, 00 or a black number different from those on which the bet was made occurring is $1 - 9/19 - 5/38 = 15/38$; in this case the amount lost is $nS + 5S = (n + 5)S$.

The probability of winning is 23/38 and probability of losing is 15/38.

The condition on *n* is $5 < n < 36$.

The next tables note the probabilities of the events attached to this experiment and the eventual afferent wins and losses for the coefficients $n = 6, 7, 8, 9, 10$.

n = 6

Event	P(/)	P(%)	Winning	Loss
winning the bet on color	9/19	47.36842	S	
winning a bet on a number	5/38	13.15789	30S	
not winning any bet	15/38	39.47368		11S

322

n = 7

Event	P(/)	P(%)	Winning	Loss
winning the bet on color	9/19	47.36842	2S	
winning a bet on a number	5/38	13.15789	29S	
not winning any bet	15/38	39.47368		12S

n = 8

Event	P(/)	P(%)	Winning	Loss
winning the bet on color	9/19	47.36842	3S	
winning a bet on a number	5/38	13.15789	28S	
not winning any bet	15/38	39.47368		13S

n = 9

Event	P(/)	P(%)	Winning	Loss
winning the bet on color	9/19	47.36842	4S	
winning a bet on a number	5/38	13.15789	27S	
not winning any bet	15/38	39.47368		14S

n = 10

Event	P(/)	P(%)	Winning	Loss
winning the bet on color	9/19	47.36842	5S	
winning a bet on a number	5/38	13.15789	26S	
not winning any bet	15/38	39.47368		15S

5. Bet amount nS on one color (red) and amount S on each of six certain numbers of the opposite color (black)

The probability of a red number occurring is 9/19; in this case the amount won is $nS - 6S = (n - 6)S$.

The probability of one of the five black numbers occurring is 3/19; in this case the amount won is $40S - nS - 5S = (35 - n)S$.

The probability of 0, 00 or a black number different from those on which the bet was made occurring is $1 - 9/19 - 3/19 = 7/19$; in this case the amount lost is $nS + 6S = (n + 6)S$.

Probability of winning is 12/19 and probability of loosing is 7/19. The condition on *n* is $6 < n < 35$.

The next tables note the probabilities of the events attached to this experiment and the eventual afferent wins and losses for the coefficients n = 7, 8, 9, 10, 11.

n = 7

Event	P(/)	P(%)	Winning	Loss
winning the bet on color	9/19	47.36842	S	
winning a bet on a number	3/19	15.78947	28S	
not winning any bet	7/19	36.84210		13S

n = 8

Event	P(/)	P(%)	Winning	Loss
winning the bet on color	9/19	47.36842	2S	
winning a bet on a number	3/19	15.78947	27S	
not winning any bet	7/19	36.84210		14S

n = 9

Event	P(/)	P(%)	Winning	Loss
winning the bet on color	9/19	47.36842	3S	
winning a bet on a number	3/19	15.78947	26S	
not winning any bet	7/19	36.84210		15S

n = 10

Event	P(/)	P(%)	Winning	Loss
winning the bet on color	9/19	47.36842	4S	
winning a bet on a number	3/19	15.78947	25S	
not winning any bet	7/19	36.84210		16S

n = 11

Event	P(/)	P(%)	Winning	Loss
winning the bet on color	9/19	47.36842	5S	
winning a bet on a number	3/19	15.78947	24S	
not winning any bet	7/19	36.84210		17S

6. Bet amount nS on one color (red) and amount S on each of seven certain numbers of the opposite color (black)

The probability of a red number occurring is 9/19; in this case the amount won is $nS - 7S = (n - 7)S$.

The probability of one of the five black numbers occurring is 7/38; in this case the amount won is $40S - nS - 6S = (34 - n)S$.

The probability of 0, 00 or a black number different from those on which the bet was made occurring is $1 - 9/19 - 7/38 = 13/38$; in this case the amount lost is $nS + 6S = (n + 6)S$.

Probability of winning is 25/38 and probability of loosing is 13/38.

The condition on n is $7 < n < 34$.

The next tables note the probabilities of the events attached to this experiment and the eventual afferent wins and losses for the coefficients $n = 8, 9, 10, 11, 12$.

$n = 8$

Event	P(/)	P(%)	Winning	Loss
winning the bet on color	9/19	47.36842	S	
winning a bet on a number	7/38	18.42105	26S	
not winning any bet	13/38	34.21052		15S

$n = 9$

Event	P(/)	P(%)	Winning	Loss
winning the bet on color	9/19	47.36842	2S	
winning a bet on a number	7/38	18.42105	25S	
not winning any bet	13/38	34.21052		16S

$n = 10$

Event	P(/)	P(%)	Winning	Loss
winning the bet on color	9/19	47.36842	3S	
winning a bet on a number	7/38	18.42105	24S	
not winning any bet	13/38	34.21052		17S

n = 11

Event	P(/)	P(%)	Winning	Loss
winning the bet on color	9/19	47.36842	4S	
winning a bet on a number	7/38	18.42105	23S	
not winning any bet	13/38	34.21052		18S

n = 12

Event	P(/)	P(%)	Winning	Loss
winning the bet on color	9/19	47.36842	5S	
winning a bet on a number	7/38	18.42105	22S	
not winning any bet	13/38	34.21052		19S

The betting combinations described in cases 1 through 6 improve the winning probability, when compared with a simple bet on a color, from 10/19 (52.63157 %) in case 1, to 25/38 (65.78947 %) in case 6.

When choosing the optimal combinations, a player must also take into account his or her personal criteria (cash amount being disposed of, playing time, etc.)

For example, in case 6 (color + 7 numbers of opposite color), n = 8, the probability of winning is good (almost 53 %), but, for a *15S* bet amount, the most probable winning amount is *S*, and this means a low winning rate and a long time before showing any profit.

By betting the entire amount *15S* on a color, for example, a player can win *15S*, with a probability of only about 47%.

But the first choice also offers a (7/38) chance of achieving a *26S* winning amount.

These betting combinations are especially recommended for low values of *S* (low stakes) because, even if the most probable winnings are not very big, they can be regular and constant.

Moreover, there is a chance of making a good profit in the case of winning a bet on a number.

The complex bets 1 through 6 can be extended to other bets composed of partially complementary events, that is: odd + even numbers, even + odd numbers, small + big numbers, big + small numbers.

All probabilities and winning data for these events will be the same as those noted in the tables presented earlier.

Another bet that is composed of partially complementary events is the bet on one of columns along with numbers not belonging to that column.

For example: bet amount nS on a column and amount S on each of three certain numbers that do not belong to that column (let us say column 1 + numbers 11, 24, 35).

The probability of winning the bet on the column is 6/19; in this case the winning amount is $2nS - 3S = (2n - 3)S$.

The probability of one of the three numbers occurring is 3/38; in this case the winning amount is $40S - nS - 2S = (38 - n)S$.

The probability of losing is $1 - 6/19 - 3/38 = 23/38$; in this case the amount lost is $nS + 3S = (n + 3)S$.

We observe that the winning probability is smaller than that corresponding to the similar case of a bet on a color (1–6).

That probability can be improved by increasing the number of bets on numbers outside the column.

In the case of the repetition of winning numbers or of their properties, we have:

The probability of a certain number occurring two consecutive times is $(1/38)(1/38) = 1/1444$ (calculated in the moment before the first spin); the probability of a certain number occurring three consecutive times is $(1/38)(1/38)(1/38) = 1/54872$, and so on.

The probability of a certain color occurring two consecutive times is $(9/19)(9/19) = 22.43767\%$ (calculated in the moment before the first spin); the probability of a certain color occurring three consecutive times is $(9/19)(9/19)(9/19) = 10.62837\%$, and so on.

The same probabilities hold for even, odd, small or big numbers to occur consecutively.

An often-practiced system, the so-called *double the bet* or *martingale* system, is based on these low probabilities of repetition and on a mathematical inequality.

The amount S is bet on a simple event (for example, the color red). If a red number does not occur in the first game, the bet is raised to $2S$ on the second game.

If a red number still does not occur, the bet is raised to *4S* in the third game, and so on.

When a red number finally occurs, the amount received will exceed the total loss from previous games, and that is algebraically provable:

Assume that *n* consecutive bets are lost and the *n +1* – th is won.

The total amount lost is $S + 2S + 4S + ... + 2^{n-1}S$, and the amount received is $2^n S$.

We have:

$$S + 2S + 4S + ... + 2^{n-1}S = S(1 + 2^2 + 2^3 + ... + 2^{n-1}) = S(2^n - 1) < 2^n S.$$

The inequality is also valid for other larger-than-2 multipliers of *S* (3, 4, etc.), and can also generate bigger winnings.

Based on this mathematical certainty and on the low probability of the consecutive repetition of color, the system seems to be infallible.

However, it assumes the major risk of consuming the entire cash amount available before the winning bet, as well as the roulette ball consistently landing on one color during a very long series of plays.

Blackjack

Blackjack is a casino card game similar to baccarat.

The object of blackjack is to accumulate cards with point totals as close to 21 as possible, without going over 21. A hand totaling 21 points is called *blackjack*.

The dealer and one or several players are in play.

Every player makes his or her own bet before each playing the hand. The bet is won if the player achieves a better score than the dealer (house).

Blackjack is played with one, two or many decks, depending on each casino's rules.

Face cards (*J, Q* and *K*) are worth 10 points. Aces are worth 1 or 11 points, at the player's option. The values of the other cards are represented by their number.

Each player is initially dealt two cards, face up. The dealer is dealt two cards, one of them face up, the other face down. This last one is shown after all players have played their hands.

Players must play their hands first, and the dealer plays his hand last if any players' hands on the table have not gone bust (any hand over 21 points is called a *bust*).

Each player asks for a card in succession until he or she receives cards having a certain total value that might beat the dealer.

The dealer must hit (take additional cards) until the dealer's hand has a point value of 17 or higher.

If the dealer has A + 6 (a *soft 17*), he must stand. If the player is dealt an ace and a 10 card (a natural blackjack), and the dealer does not have blackjack, the player is immediately paid 1.5 times his bet (3 to 2). If the dealer is dealt a blackjack and is not showing an ace, the player loses the bet.

Insurance: Whenever the dealer's up card is an ace, an insurance option is available. When buying insurance, the player is insuring himself against the possibility that the dealer has a blackjack.

Insurance is an additional wager equal to half of the original bet.

If the dealer has a blackjack, the insurance pays off at 2:1.

The player's original bet is lost, unless the player also has a blackjack, in which case the hand is a *push*.

If the dealer does not have a blackjack, player loses the insurance wager and the hand is played out as usual.

Split: If player is dealt two cards of the same value (8–8, A–A, 10–K, etc.), he or she may choose to split the cards into two separate hands. An additional wager equal to the original wager is placed next to the additional hand.

The dealer then deals a second card to the first of the split hands. This hand is played out with all the normal options available.

When the player is finished playing the first hand, the other split hand is then dealt a second card and played out as usual.

A split-hand blackjack pays off at 1:1 instead of 3:2.

Double Down: If a player feels that the chances of winning a hand are exceptionally good, he or she can choose to double the amount of the original wager.

The player is dealt exactly one additional card and play passes to the dealer.

Surrender: If a player feels that the chances of winning a hand are not good, he or she has the option to surrender the hand.

If the player chooses to surrender, he or she automatically loses half of the original wager; the other half is returned to player; play ceases and the deck is reshuffled.

However, if the dealer is dealt a blackjack, the option to surrender is not available.

Depending on the casino, the game might have additional or different rules.

We first present the probabilities attached to card dealing and initial predictions.

In making this calculus, circumstantial information such as fraudulent dealing is not taken into account (as in all situations corresponding to card games).

All probabilities are calculated for cases using one or two decks of cards.

Let us look at the probabilities for a favorable initial hand (the first two cards dealt) to be achieved.

The total number of possible combinations for each of the two cards is $C_{52}^2 = 1326$, for the 1-deck game and $C_{104}^2 = 5356$ for the 2-deck game.

Probability of obtaining a natural blackjack

a) For one deck

The favorable variants are those containing an ace and a 10 (A + 10). Their number is 16 (4 aces x 4 10's), so the probability is $P = 16/1326 = 8/663$.

b) For two decks

The number of favorable variants is 64 (8 aces x 8 10's). The probability is $P = 64/5356 = 16/1339$.

Probability of obtaining a natural blackjack is $P = 8/663 = 1.20663\%$ in the case of a 1-deck game and $P = 16/1339 = 1.19492\%$ in the case of a 2-deck game.

Probability of obtaining a blackjack

a) For one deck

The favorable variants are those of the A + 10 type (ace + card with the value 10) and their number is 64 (4 aces x 16 cards with the value 10).

Therefore, the probability is $P = 64/1326 = 32/663$.

b) For two decks

The number of favorable variants is 256 (8 aces x 32 cards with the value 10), so the probability is $P = 256/5356 = 64/1339$.

Probability of obtaining a blackjack from the first two cards is $P = 32/663 = 4.82654\%$ in the case of a 1-deck game and $P = 64/1339 = 4.77968\%$ in the case of a 2-deck game.

Similarly, we can calculate the following probabilities:

Probability of obtaining 20 points from the first two cards is $P = 68/663 = 10.25641\%$ in the case of a 1-deck game and $P = 140/1339 = 10.45556\%$ in the case of a 2-deck game.

Probability of obtaining 19 points from the first two cards is $P = 40/663 = 6.03318\%$ in the case of a 1-deck game and $P = 80/1339 = 5.97460\%$ in the case of a 2-deck game.

Probability of obtaining 18 points from the first two cards is $P = 43/663 = 6.48567\%$ in the case of a 1-deck game and $P = 87/1339 = 6.4973\%$ in the case of a 2-deck game.

Probability of getting 17 points from the first two cards is
P = 16/221 = 7.23981% **in the case of a 1-deck game and**
P = 96/1339 = 7.16952% **in the case of a 2-deck game.**

As you can see, the probabilities are approximately equal for both cases, whether one deck or two decks are used.

A *good* initial hand (which you can stay with) could be a blackjack or a hand of 20, 19 or 18 points.

The probability of obtaining such a hand is calculated by totaling the corresponding probabilities calculated above:

P = 32/663 + 68/663 + 40/663 + 43/663 = 183/663, in the case of a 1-deck game and

P = 64/1339 + 140/1339 + 80/1339 + 87/1339 = 371/1339, in the case of a 2-deck game.

Probability of obtaining a *good* initial hand is
P = 183/663 = 27.60180% **in the case of a 1-deck game and**
P = 371/1339 = 27.70724% **in the case of a 2-deck game.**

The probabilities of events predicted during the game are calculated on the basis of the played cards (the cards showing) from a certain moment.

This requires counting certain favorable cards showing for the dealer and for the other players, as well as in your own hand.

Any blackjack strategy is based on counting the cards played.

Unlike a baccarat game, where a maximum of three cards are played for each player, at blackjack many cards could be played at a certain moment, especially when many players are at the table.

Thus, both following and memorizing certain cards require some ability and prior training on the player's part.

The formula of probability for obtaining a certain favorable value is similar to that for baccarat and depends on the number of decks of cards used.

If we denote by x a favorable value, by nx the number of cards showing with the value x (from your hand, the hands of the other players and the face up card in the dealer's hand) and by nv the total number of cards showing, then the probability of the next card from the deck (the one you receive if you ask for an additional card) having the value x is:

$$P = \frac{4 - nx}{52 - nv}, \text{ if } x \neq 10 \text{ and}$$

$$P = \frac{16 - nx}{52 - nv}, \text{ if } x = 10.$$

This formula holds for the case of a 1-deck game.
In the case of a 2-deck game, the probability is:

$$P = \frac{8 - nx}{104 - nv}, \text{ if } x \neq 10 \text{ and}$$

$$P = \frac{32 - nx}{104 - nv}, \text{ if } x = 10.$$

Generally speaking, if playing with m decks, the probability of obtaining a card with the value x is:

$$P = \frac{4m - nx}{52m - nv}, \text{ if } x \neq 10 \text{ and}$$

$$P = \frac{16m - nx}{52m - nv}, \text{ if } x = 10.$$

Examples of application of the formula:

1) Assume play with one deck, you are the only player at table, you hold Q, 2, 4, A (total value 17) and the face up card of the dealer is a 4.

Let us calculate the probability of achieving 21 points (receiving a 4).

We have $nx = 2$, $nv = 5$, so:

$$P = \frac{4 - 2}{52 - 5} = 2 / 47 = 4.25531\%.$$

For the probability of achieving 20 points (receiving a 3), we have $nx = 0$, $nv = 5$, so:

$$P = \frac{4 - 0}{52 - 5} = 4 / 47 = 8.51063\%.$$

For the probability of achieving 19 points (receiving a 2), we have $nx = 1$, $nv = 5$, so:

$$P = \frac{4-1}{52-5} = 3/47 = 6.38297\%.$$

If we want to calculate the probability of achieving 19, 20 or 21 points, all we must do is total the three probabilities just calculated. We obtain $P = 9/47 = 19.14893\%$.

2) Assume play with three decks, you are the third player in play, the face up card of the dealer is a 7 and the previous players have the following hands:

the first: 5, 3, J, 8,

the second: A, 6, 2, 3, 7, A,

You hold: 5, 7, 4, 3.

Let us calculate the probability of achieving 20 points (receiving an ace).

We have $nx = 2$, $nv = 15$, so:

$$P = \frac{4 \cdot 3 - 2}{52 \cdot 3 - 15} = 10/141 = 7.09219\%.$$

Let us calculate the probability of achieving 21 points (receiving a 2).

We have $nx = 1$, $nv = 15$, so:

$$P = \frac{4 \cdot 3 - 1}{52 \cdot 3 - 15} = 11/141 = 7.80141\%.$$

The probability of receiving a favorable value (A or 2) is $10/141 + 11/141 = 21/141 = 14.89361\%$.

From the formulas' expressions we can immediately deduce that the probability will be higher as nx becomes lower and nv becomes larger.

The probability is maximum for $nx = 0$ (the situation in which no cards can be seen).

Obviously, $nv \geq 3$ (one dealer's card + your own two initial cards).

We cannot establish an upper limit for nv because this limit depends directly on the number of players and their order in the game; *nv* is maximum when the number of players is maximum and you are the last in play.

Let us make an arbitrary estimation. Assume play with one deck, two players are at the table, each has seven cards showing, you are the last in play and you hold five cards.

So, nv = 1 + 7 + 7 + 5 = 20.

If nx = 0, the probability of receiving a favorable card with the value x is $P = \dfrac{4-0}{52-20} = 1/8$ and this is a small enough probability.

Assuming there are two other favorable values with nx = 0, the total probability increases to 3/8.

However, this calculus has been performed in highly favorable conditions for you (many cards already played, no favorable card played) which shows that the intermediary probabilities involved in the game of blackjack are mostly low.

Because of this, the decision to ask for an additional card based on an evaluation of probability is an optimum strategy.

The general algorithm for applying such a strategy could be the following:

1. Ask for an additional card until you reach a total value as close as possible to 17 (15, 16, 17). If you do not exceed 21, go to the next step.

2. Calculate the necessary favorable values. (For example, if you hold a total of 16 points, the favorable values are calculated by subtracting 16 from 21, 20, 19, 18 or 17: 21 – 16 = 5, 20 – 16 = 4, 19 – 16 = 3, 18 – 16 = 2, 17 – 16 = 1. The favorable values are 1, 2, 3, 4, 5).

3. View all face up cards, including those in your hand. Count how many cards are already played for each favorable value calculated at step 2 and memorize these numbers (*nx*).

4. Count all cards that are already played (*nv*).

5. For each favorable value, calculate the probability, according to the formula (do not reduce the fraction).

6. Memorize the results (from step 5) and total them (that is, total the fractions' numerators), then make a decimal approximation.

7. Decision: If the total probability obtained in the previous step is satisfactory, ask for an additional card. If not, stop.

8. If you do not exceed 21, re-evaluate the probability of receiving favorable values by repeating steps 2 through 7 for the new situation.

At step 7, establishing a threshold between *satisfactory* and *unsatisfactory* is a matter exclusively of the player's competence, the risk level assumed and other personal criteria.

Unlike in baccarat, where fewer cards are played, the number of players is constant (two), and the number of gaming situations is very limited, in blackjack, the number of possible playing configurations is in the thousands and, as a practical matter, cannot be covered by tables of values.

However, we still present the precalculated probabilities for this section. They are the probabilities attached to gaming events from the following situations (frequently met): you are the only player at the table or there are several players and you are the first in play.

For the situation in which you are the only player, the probabilities are calculated for a 1-deck game.

We now consider gaming situations that require a decision, where the total value held is 15, 16, 17, 18, 19 or 20 points.

For a total value of less than 15, the probability that the dealer wins is quite large (the dealer must stay at a minimum of 17; therefore, the case in which you win is practically equivalent to the dealer exceeding 21).

At the same time, we consider as favorable those values which, by totaling the total value held, give the sum 17, 18, 19, 20 or 21.

These represent a total value you can stay with while waiting for the dealer to play. Thus:

– for a total value of 15, the favorable values are 2, 3, 4, 5, 6
– for a total value of 16, the favorable values are 1, 2, 3, 4, 5
– for a total value of 17, the favorable values are 1, 2, 3, 4
– for a total value of 18, the favorable values are 1, 2, 3
– for a total value of 19, the favorable values are 1, 2
– for a total value of 20, the favorable value is 1.

The total number of variants in the distribution of cards for a hand holding a total value of 15, 16, 17, 18, 19 or 20 is in the thousands, so they cannot be presented in their entirety.

We do consider the variants in distribution for hands of two and three cards, particularly situations in which you hold two or three cards having a total value between 15 and 20.

The following tables note the probabilities of receiving each favorable value and also the total probabilities for all the cases described above (two or three cards in a hand).

We remind the reader that the probabilities from these tables are calculated for the situation in which you are the only player at the table and the game is played with one deck of cards (as it turns out, for a two-deck game, the probabilities are not much different).

Each row corresponds to one variant in card distribution in your hand. The columns correspond to favorable values.

For each favorable value, the following cases are noted:

A – the dealer's visible card has no particular favorable value;

B – the dealer's visible card has a favorable value.

For the total probability, the following cases are noted:

C – the dealer's visible card has no particular favorable value;

D – the dealer's visible card has one of the favorable values.

total value 15 (favorable values: 2, 3, 4, 5, 6)

Variant	Favorable values										Total probability	
	2		3		4		5		6			
	A	B	A	B	A	B	A	B	A	B	C	D
11 + 4	4/49 8.16326 %	3/49 6.12244 %	4/49 8.16326 %	3/49 6.12244 %	3/49 6.12244 %	2/49 4.08163%	4/49 8.16326 %	3/49 6.12244 %	4/49 8.16326 %	3/49 6.12244 %	19/49 38.77551 %	18/49 36.73469 %
10 + 5	4/49 8.16326 %	3/49 6.12244 %	4/49 8.16326 %	3/49 6.12244 %	4/49 8.16326 %	3/49 6.12244 %	3/49 6.12244 %	2/49 4.08163 %	4/49 8.16326 %	3/49 6.12244 %	19/49 38.77551 %	18/49 36.73469 %
9 + 6	4/49 8.16326 %	3/49 6.12244 %	4/49 8.16326 %	3/49 6.12244 %	4/49 8.16326 %	3/49 6.12244 %	3/49 6.12244 %	2/49 4.08163 %	3/49 6.12244 %	2/49 4.08163 %	19/49 38.77551 %	18/49 36.73469 %
8 + 7	4/49 8.16326 %	3/49 6.12244 %	4/49 8.16326 %	3/49 6.12244 %	4/49 8.16326 %	3/49 6.12244 %	3/49 6.12244 %	2/49 4.08163 %	3/49 6.12244 %	2/49 4.08163 %	20/49 40.81632 %	19/49 38.77551 %
11 + 3 + 1	1/12 8.33333 %	1/16 6.25 %	1/16 6.25 %	1/24 4.16666 %	1/12 8.33333 %	1/16 6.25 %	1/12 8.33333 %	1/16 6.25 %	1/12 8.33333 %	1/16 6.25 %	19/48 39.58333 %	3/8 37.5 %
11 + 2 + 2	1/24 4.16666 %	1/48 2.08333 %	1/12 8.33333 %	1/16 6.25 %	1/12 8.33333 %	1/16 6.25 %	1/12 8.33333 %	1/16 6.25 %	1/12 8.33333 %	1/16 6.25 %	3/8 37.5 %	17/48 35.41666 %
10 + 4 + 1	1/12 8.33333 %	1/16 6.25 %	1/12 8.33333 %	1/16 6.25 %	1/16 6.25 %	1/24 4.16666 %	1/12 8.33333 %	1/16 6.25 %	1/12 8.33333 %	1/16 6.25 %	19/48 39.58333 %	3/8 37.5 %
10 + 3 + 2	1/16 6.25 %	1/24 4.16666 %	1/12 8.33333 %	1/16 6.25 %	1/12 8.33333 %	1/16 6.25 %	1/12 8.33333 %	1/16 6.25 %	1/12 8.33333 %	1/16 6.25 %	3/8 37.5 %	17/48 35.41666 %
9 + 5 + 1	1/12 8.33333 %	1/16 6.25 %	1/12 8.33333 %	1/16 6.25 %	1/12 8.33333 %	1/16 6.25 %	1/16 6.25 %	1/24 4.16666 %	1/12 8.33333 %	1/16 6.25 %	19/48 39.58333 %	3/8 37.5 %
9 + 4 + 2	1/16 6.25 %	1/24 4.16666 %	1/12 8.33333 %	1/16 6.25 %	1/16 6.25 %	1/24 4.16666 %	1/12 8.33333 %	1/16 6.25 %	1/12 8.33333 %	1/16 6.25 %	3/8 37.5 %	17/48 35.41666 %

continuation: total value 15 (favorable values: 2, 3, 4, 5, 6)

Variant	2 A	2 B	3 A	3 B	4 A	4 B	5 A	5 B	6 A	6 B	Total probability C	Total probability D
9+3+3	1/12 8.33333%	1/16 6.25%	1/24 4.16666%	1/48 2.08333%	1/12 8.33333%	1/16 6.25%	1/12 8.33333%	1/16 6.25%	1/12 8.33333%	1/16 6.25%	19/48 39.58333%	3/8 37.5%
8+6+1	1/12 8.33333%	1/16 6.25%	1/12 8.33333%	1/16 6.25%	1/12 8.33333%	1/16 6.25%	1/12 8.33333%	1/16 6.25%	1/16 6.25%	1/24 4.16666%	3/8 37.5%	17/48 35.41666%
8+5+2	1/16 6.25%	1/16 6.25%	1/12 8.33333%	1/16 6.25%	1/12 8.33333%	1/16 6.25%	1/16 6.25%	1/24 4.16666%	1/12 8.33333%	1/24 4.16666%	3/8 37.5%	17/48 35.41666%
8+4+3	1/12 8.33333%	1/24 4.16666%	1/16 6.25%	1/24 4.16666%	1/16 6.25%	1/24 4.16666%	1/12 8.33333%	1/16 6.25%	1/12 8.33333%	1/16 6.25%	3/8 37.5%	17/48 35.41666%
7+7+1	1/12 8.33333%	1/16 6.25%	1/12 8.33333%	1/16 6.25%	1/12 8.33333%	1/24 4.16666%	1/12 8.33333%	1/16 6.25%	1/12 8.33333%	1/16 6.25%	5/12 41.66666%	19/48 39.58333%
7+6+2	1/16 6.25%	1/24 4.16666%	1/12 8.33333%	1/16 6.25%	1/12 8.33333%	1/16 6.25%	1/12 8.33333%	1/16 6.25%	1/12 8.33333%	1/16 6.25%	3/8 37.5%	17/48 35.41666%
7+5+3	1/12 8.33333%	1/16 6.25%	1/12 8.33333%	1/24 4.16666%	1/12 8.33333%	1/16 6.25%	1/12 8.33333%	1/24 4.16666%	1/12 8.33333%	1/16 6.25%	3/8 37.5%	17/48 35.41666%
7+4+4	1/12 8.33333%	1/16 6.25%	1/16 6.25%	1/24 4.16666%	1/24 4.16666%	1/16 6.25%	1/16 6.25%	1/16 6.25%	1/24 4.16666%	1/24 4.16666%	17/48 35.41666%	17/48 35.41666%
6+6+3	1/12 8.33333%	1/16 6.25%	1/12 8.33333%	1/16 6.25%	1/12 8.33333%	1/48 2.08333%	1/12 8.33333%	1/16 6.25%	1/12 8.33333%	1/48 2.08333%	17/48 35.41666%	1/3 33.33333%
6+5+4	1/12 8.33333%	1/16 6.25%	1/12 8.33333%	1/16 6.25%	1/16 6.25%	1/24 4.16666%	1/16 6.25%	1/24 4.16666%	1/16 6.25%	1/24 4.16666%	17/48 35.41666%	1/3 33.33333%
5+5+5	1/12 8.33333%	1/16 6.25%	1/12 8.33333%	1/16 6.25%	1/12 8.33333%	1/16 6.25%	1/48 2.08333%	0	1/12 8.33333%	1/16 6.25%	17/48 35.41666%	1/3 33.33333%

total value 16 (favorable values: 1, 2, 3, 4, 5)

Variant	1 A	1 B	2 A	2 B	3 A	3 B	4 A	4 B	5 A	5 B	Total probability C	Total probability D
11 + 5	3/49 6.12244 %	2/49 4.08163 %	4/49 8.16326 %	3/49 6.12244 %	4/49 8.16326 %	3/49 6.12244 %	4/49 8.16326 %	3/49 6.12244 %	3/49 6.12244 %	2/49 4.08163 %	18/49 36.73469 %	17/49 34.69387 %
10 + 6	4/49 8.16326 %	3/49 6.12244 %	4/49 8.16326 %	3/49 6.12244 %	4/49 8.16326 %	3/49 6.12244 %	4/49 8.16326 %	3/49 6.12244 %	4/49 8.16326 %	3/49 6.12244 %	20/49 40.81632 %	19/49 38.77551 %
9 + 7	4/49 8.16326 %	3/49 6.12244 %	4/49 8.16326 %	3/49 6.12244 %	4/49 8.16326 %	3/49 6.12244 %	4/49 8.16326 %	3/49 6.12244 %	4/49 8.16326 %	3/49 6.12244 %	20/49 40.81632 %	19/49 38.77551 %
8 + 8	4/49 8.16326 %	3/49 6.12244 %	4/49 8.16326 %	3/49 6.12244 %	4/49 8.16326 %	3/49 6.12244 %	4/49 8.16326 %	3/49 6.12244 %	4/49 8.16326 %	3/49 6.12244 %	20/49 40.81632 %	19/49 38.77551 %
11 + 4 + 1	1/24 4.16666 %	1/48 2.08333 %	1/12 8.33333 %	1/16 6.25 %	1/12 8.33333 %	1/16 6.25 %	1/12 8.33333 %	1/24 4.16666 %	1/12 8.33333 %	1/16 6.25 %	17/48 35.41666 %	1/3 33.33333 %
11 + 3 + 2	1/16 6.25 %	1/24 4.16666 %	1/16 6.25 %	1/24 4.16666 %	1/16 6.25 %	1/24 4.16666 %	1/12 8.33333 %	1/16 6.25 %	1/12 8.33333 %	1/16 6.25 %	17/48 35.41666 %	1/3 33.33333 %
10 + 5 + 1	1/16 6.25 %	1/24 4.16666 %	1/12 8.33333 %	1/16 6.25 %	1/12 8.33333 %	1/16 6.25 %	1/16 6.25 %	1/16 6.25 %	1/16 6.25 %	1/24 4.16666 %	3/8 37.5 %	17/48 35.41666 %
10 + 4 + 2	1/12 8.33333 %	1/16 6.25 %	1/12 8.33333 %	1/16 6.25 %	1/24 4.16666 %	1/48 2.08333 %	1/16 6.25 %	1/24 4.16666 %	1/12 8.33333 %	1/16 6.25 %	3/8 37.5 %	17/48 35.41666 %
10 + 3 + 3	1/12 8.33333 %	1/16 6.25 %	1/12 8.33333 %	1/16 6.25 %	1/12 8.33333 %	1/16 6.25 %	1/12 8.33333 %	1/16 6.25 %	1/12 8.33333 %	1/16 6.25 %	3/8 37.5 %	17/48 35.41666 %
9 + 6 + 1	1/16 6.25 %	1/24 4.16666 %	1/12 8.33333 %	1/16 6.25 %	1/12 8.33333 %	1/16 6.25 %	1/12 8.33333 %	1/16 6.25 %	1/12 8.33333 %	1/16 6.25 %	19/48 39.58333 %	3/8 37.5 %

continuation: total value 16 (favorable values: 1, 2, 3, 4, 5)

Variant	1 A	1 B	2 A	2 B	3 A	3 B	4 A	4 B	5 A	5 B	Total prob. C	Total prob. D
9 + 5 + 2	1/12 8.33333%	1/16 6.25%	1/16 6.25%	1/24 4.16666%	1/12 8.33333%	1/16 6.25%	1/12 8.33333%	1/16 6.25%	1/16 6.25%	1/24 4.16666%	3/8 37.5%	17/48 35.41666%
9 + 4 + 3	1/12 8.33333%	1/16 6.25%	1/12 8.33333%	1/16 6.25%	1/16 6.25%	1/24 4.16666%	1/16 6.25%	1/24 4.16666%	1/12 8.33333%	1/16 6.25%	3/8 37.5%	17/48 35.41666%
8 + 7 + 1	1/16 6.25%	1/24 4.16666%	1/12 8.33333%	1/16 6.25%	1/12 8.33333%	1/16 6.25%	1/12 8.33333%	1/16 6.25%	1/12 8.33333%	1/16 6.25%	19/48 39.58333%	3/8 37.5%
8 + 8 + 2	1/12 8.33333%	1/16 6.25%	1/12 8.33333%	1/24 4.16666%	1/12 8.33333%	1/16 6.25%	1/12 8.33333%	1/16 6.25%	1/12 8.33333%	1/16 6.25%	19/48 39.58333%	3/8 37.5%
8 + 5 + 3	1/12 8.33333%	1/16 6.25%	1/16 6.25%	1/24 4.16666%	1/16 6.25%	1/24 4.16666%	1/12 8.33333%	1/16 6.25%	1/16 6.25%	1/16 6.25%	3/8 37.5%	17/48 35.41666%
8 + 4 + 4	1/12 8.33333%	1/16 6.25%	1/12 8.33333%	1/16 6.25%	1/12 8.33333%	1/16 6.25%	1/24 4.16666%	1/48 2.08333%	1/12 8.33333%	1/16 6.25%	3/8 37.5%	17/48 35.41666%
7 + 7 + 2	1/12 8.33333%	1/16 6.25%	1/16 6.25%	1/16 6.25%	1/12 8.33333%	1/16 6.25%	1/12 8.33333%	1/16 6.25%	1/12 8.33333%	1/16 6.25%	3/8 37.5%	17/48 35.41666%
7 + 6 + 3	1/12 8.33333%	1/16 6.25%	1/16 6.25%	1/24 4.16666%	1/12 8.33333%	1/24 4.16666%	1/12 8.33333%	1/16 6.25%	1/12 8.33333%	1/16 6.25%	19/48 39.58333%	3/8 37.5%
7 + 5 + 4	1/12 8.33333%	1/16 6.25%	1/12 8.33333%	1/16 6.25%	1/16 6.25%	1/16 6.25%	1/16 6.25%	1/24 4.16666%	1/12 8.33333%	1/16 6.25%	19/48 39.58333%	3/8 37.5%
6 + 6 + 4	1/12 8.33333%	1/16 6.25%	1/12 8.33333%	1/16 6.25%	1/12 8.33333%	1/16 6.25%	1/16 6.25%	1/24 4.16666%	1/16 6.25%	1/24 4.16666%	3/8 37.5%	3/8 37.5%
6 + 5 + 5	1/12 8.33333%	1/16 6.25%	1/12 8.33333%	1/16 6.25%	1/12 8.33333%	1/16 6.25%	1/12 8.33333%	1/16 6.25%	1/16 6.25%	1/24 4.16666%	3/8 37.5%	17/48 35.41666%

total value 17 (favorable values: 1, 2, 3, 4)

Variant	Favorable values								Total probability	
	1		2		3		4			
	A	B	A	B	A	B	A	B	C	D
11+6	3/49 6.12244%	2/49 4.08163%	4/49 8.16326%	3/49 6.12244%	4/49 8.16326%	3/49 6.12244%	4/49 8.16326%	3/49 6.12244%	15/49 30.61224%	2/7 28.57142%
10+7	4/49 8.16326%	3/49 6.12244%	4/49 8.16326%	3/49 6.12244%	4/49 8.16326%	3/49 6.12244%	4/49 8.16326%	3/49 6.12244%	16/49 32.65306%	15/49 30.61224%
9+8	4/49 8.16326%	3/49 6.12244%	4/49 8.16326%	3/49 6.12244%	4/49 8.16326%	3/49 6.12244%	4/49 8.16326%	3/49 6.12244%	16/49 32.65306%	15/49 30.61224%
11+5+1	1/24 4.16666%	1/48 2.08333%	1/12 8.33333%	1/16 6.25%	1/12 8.33333%	1/16 6.25%	1/12 8.33333%	1/16 6.25%	7/24 29.16666%	13/48 27.08333%
11+4+2	1/16 6.25%	1/24 4.16666%	1/16 6.25%	1/24 4.16666%	1/12 8.33333%	1/16 6.25%	1/16 6.25%	1/24 4.16666%	13/48 27.08333%	1/4 25%
11+3+3	1/16 6.25%	1/24 4.16666%	1/12 8.33333%	1/16 6.25%	1/24 4.16666%	1/48 2.08333%	1/12 8.33333%	1/16 6.25%	13/48 27.08333%	1/4 25%
10+6+1	1/16 6.25%	1/24 4.16666%	1/12 8.33333%	1/16 6.25%	1/12 8.33333%	1/16 6.25%	1/12 8.33333%	1/16 6.25%	5/16 31.25%	7/24 29.16666%
10+5+2	1/12 8.33333%	1/16 6.25%	1/16 6.25%	1/16 6.25%	1/12 8.33333%	1/16 6.25%	1/12 8.33333%	1/16 6.25%	5/16 31.25%	7/24 29.16666%
10+4+3	1/12 8.33333%	1/16 6.25%	1/12 8.33333%	1/16 6.25%	1/16 6.25%	1/24 4.16666%	1/16 6.25%	1/24 4.16666%	7/24 29.16666%	13/48 27.08333%
9+7+1	1/16 6.25%	1/24 4.16666%	1/12 8.33333%	1/16 6.25%	1/12 8.33333%	1/16 6.25%	1/12 8.33333%	1/16 6.25%	5/16 31.25%	7/24 29.16666%

342

continuation: total value 17 (favorable values: 1, 2, 3, 4)

Variant	Favorable values								Total probability	
	1		**2**		**3**		**4**			
	A	B	A	B	A	B	A	B	C	D
9+6+2	1/12 8.33333%	1/16 6.25%	1/16 6.25%	1/24 4.16666%	1/12 8.33333%	1/16 6.25%	1/12 8.33333%	1/16 6.25%	5/16 31.25%	7/24 29.16666%
9+5+3	1/12 8.33333%	1/16 6.25%	1/12 8.33333%	1/16 6.25%	1/16 6.25%	1/24 4.16666%	1/12 8.33333%	1/16 6.25%	5/16 31.25%	7/24 29.16666%
9+4+4	1/12 8.33333%	1/24 4.16666%	1/12 8.33333%	1/16 6.25%	1/12 8.33333%	1/16 6.25%	1/24 4.16666%	1/48 2.08333%	7/24 29.16666%	13/48 27.08333%
8+8+1	1/16 6.25%	1/16 6.25%	1/16 6.25%	1/16 6.25%	1/12 8.33333%	1/16 6.25%	1/12 8.33333%	1/16 6.25%	5/16 31.25%	7/24 29.16666%
8+7+2	1/12 8.33333%	1/16 6.25%	1/12 8.33333%	1/24 4.16666%	1/12 8.33333%	1/16 6.25%	1/12 8.33333%	1/16 6.25%	5/16 31.25%	7/24 29.16666%
8+6+3	1/12 8.33333%	1/16 6.25%	1/12 8.33333%	1/16 6.25%	1/16 6.25%	1/24 4.16666%	1/12 8.33333%	1/24 4.16666%	5/16 31.25%	7/24 29.16666%
8+5+4	1/12 8.33333%	1/16 6.25%	1/12 8.33333%	1/16 6.25%	1/12 8.33333%	1/16 6.25%	1/16 6.25%	1/16 6.25%	5/16 31.25%	7/24 29.16666%
7+7+3	1/12 8.33333%	1/16 6.25%	1/12 8.33333%	1/16 6.25%	1/16 6.25%	1/24 4.16666%	1/12 8.33333%	1/16 6.25%	5/16 31.25%	7/24 29.16666%
7+6+4	1/12 8.33333%	1/16 6.25%	1/12 8.33333%	1/16 6.25%	1/12 8.33333%	1/16 6.25%	1/12 8.33333%	1/16 6.25%	5/16 31.25%	7/24 29.16666%
7+5+5	1/12 8.33333%	1/16 6.25%	1/12 8.33333%	1/16 6.25%	1/12 8.33333%	1/12 8.33333%	1/12 8.33333%	1/16 6.25%	1/3 33.333%	1/4 25%
6+6+5	1/12 8.33333%	1/16 6.25%	1/12 8.33333%	1/16 6.25%	1/12 8.33333%	1/16 6.25%	1/12 8.33333%	1/16 6.25%	1/3 33.333%	1/4 25%

total value 18 (favorable values: 1, 2, 3)

Variant	Favorable values						Total probability	
	1		2		3			
	A	B	A	B	A	B	C	D
11 + 7	3/49 6.12244 %	2/49 4.08163 %	4/49 8.16326 %	3/49 6.12244 %	4/49 8.16326 %	3/49 6.12244 %	11/49 22.44897%	10/49 20.40816%
10 + 8	4/49 8.16326 %	3/49 6.12244 %	4/49 8.16326 %	3/49 6.12244 %	4/49 8.16326 %	3/49 6.12244 %	12/49 24.48979%	11/49 22.44897%
9 + 9	4/49 8.16326 %	3/49 6.12244 %	4/49 8.16326 %	3/49 6.12244 %	4/49 8.16326 %	3/49 6.12244 %	12/49 24.48979%	11/49 22.44897%
11 + 6 + 1	1/24 4.16666 %	1/48 2.08333 %	1/12 8.33333 %	1/16 6.25 %	1/12 8.33333 %	1/16 6.25 %	5/24 20.83333%	3/16 18.75 %
11 + 5 + 2	1/16 6.25 %	1/24 4.16666 %	1/16 6.25 %	1/24 4.16666 %	1/16 6.25 %	1/24 4.16666 %	5/24 20.83333%	3/16 18.75 %
11 + 4 + 3	1/16 6.25 %	1/24 4.16666 %	1/12 8.33333 %	1/16 6.25 %	1/16 6.25 %	1/24 4.16666 %	11/48 22.91666%	5/24 20.83333%
10 + 7 + 1	1/16 6.25 %	1/24 4.16666 %	1/12 8.33333 %	1/16 6.25 %	1/12 8.33333 %	1/16 6.25 %	11/48 22.91666%	5/24 20.83333%
10 + 6 + 2	1/12 8.33333 %	1/16 6.25 %	1/12 8.33333 %	1/24 4.16666 %	1/12 8.33333 %	1/16 6.25 %	11/48 22.91666%	5/24 20.83333%
10 + 5 + 3	1/12 8.33333 %	1/16 6.25 %	1/12 8.33333 %	1/16 6.25 %	1/16 6.25 %	1/24 4.16666 %	11/48 22.91666%	5/24 20.83333%
10 + 4 + 4	1/12 8.33333 %	1/16 6.25 %	1/12 8.33333 %	1/16 6.25 %	1/12 8.33333 %	1/16 6.25 %	1/4 25 %	11/48 22.91666%

continuation: total value 18 (favorable values: 1, 2, 3)

Variant	Favorable values						Total probability	
	1		2		3			
	A	B	A	B	A	B	C	D
9 + 8 + 1	1/16 6.25 %	1/24 4.16666 %	1/12 8.33333 %	1/16 6.25 %	1/12 8.33333 %	1/16 6.25 %	11/48 22.91666%	5/24 20.83333%
9 + 7 + 2	1/12 8.33333 %	1/16 6.25 %	1/16 6.25 %	1/24 4.16666 %	1/12 8.33333 %	1/16 6.25 %	11/48 22.91666%	5/24 20.83333%
9 + 6 + 3	1/12 8.33333 %	1/16 6.25 %	1/12 8.33333 %	1/16 6.25 %	1/16 6.25 %	1/24 4.16666 %	11/48 22.91666%	5/24 20.83333%
9 + 5 + 4	1/12 8.33333 %	1/16 6.25 %	1/12 8.33333 %	1/16 6.25 %	1/12 8.33333 %	1/16 6.25 %	1/4 25 %	11/48 22.91666%
8 + 8 + 2	1/12 8.33333 %	1/16 6.25 %	1/16 6.25 %	1/24 4.16666 %	1/12 8.33333 %	1/16 6.25 %	11/48 22.91666%	5/24 20.83333%
8 + 7 + 3	1/12 8.33333 %	1/16 6.25 %	1/12 8.33333 %	1/16 6.25 %	1/16 6.25 %	1/24 4.16666 %	11/48 22.91666%	5/24 20.83333%
8 + 6 + 4	1/12 8.33333 %	1/16 6.25 %	1/12 8.33333 %	1/16 6.25 %	1/12 8.33333 %	1/16 6.25 %	1/4 25 %	11/48 22.91666%
8 + 5 + 5	1/12 8.33333 %	1/16 6.25 %	1/12 8.33333 %	1/16 6.25 %	1/12 8.33333 %	1/16 6.25 %	1/4 25 %	11/48 22.91666%
7 + 7 + 4	1/12 8.33333 %	1/16 6.25 %	1/12 8.33333 %	1/16 6.25 %	1/12 8.33333 %	1/16 6.25 %	1/4 25 %	11/48 22.91666%
7 + 6 + 5	1/12 8.33333 %	1/16 6.25 %	1/12 8.33333 %	1/16 6.25 %	1/12 8.33333 %	1/12 8.33333 %	1/4 25 %	11/48 22.91666%
6 + 6 + 6	1/12 8.33333 %	1/16 6.25 %	1/12 8.33333 %	1/16 6.25 %	1/12 8.33333 %	1/16 6.25 %	1/4 25 %	11/48 22.91666%

total value 19 (favorable values: 1, 2)

Variant	Favorable values				Total probability	
	1		2			
	A	B	A	B	C	D
11 + 8	4/49 8.16326 %	3/49 6.12244 %	4/49 8.16326 %	3/49 6.12244 %	8/49 16.3265%	1/7 14.28571%
10 + 9	4/49 8.16326 %	3/49 6.12244 %	4/49 8.16326 %	3/49 6.12244 %	8/49 16.3265%	1/7 14.28571%
11 + 7 + 1	1/24 4.16666 %	1/48 2.08333 %	1/12 8.33333 %	1/16 6.25 %	1/8 12.5 %	5/48 10.41666%
11 + 6 + 2	1/16 6.25 %	1/24 4.16666 %	1/16 6.25 %	1/24 4.16666 %	1/8 12.5 %	5/48 10.41666%
11 + 5 + 3	1/16 6.25 %	1/24 4.16666 %	1/12 8.33333 %	1/16 6.25 %	7/48 14.5833%	1/8 12.5 %
11 + 4 + 4	1/16 6.25 %	1/24 4.16666 %	1/12 8.33333 %	1/16 6.25 %	7/48 14.5833%	1/8 12.5 %
10 + 8 + 1	1/16 6.25 %	1/24 4.16666 %	1/12 8.33333 %	1/16 6.25 %	7/48 14.5833%	1/8 12.5 %
10 + 7 + 2	1/12 8.33333 %	1/16 6.25 %	1/16 6.25 %	1/24 4.16666 %	7/48 14.5833%	1/8 12.5 %
10 + 6 + 3	1/12 8.33333 %	1/16 6.25 %	1/12 8.33333 %	1/16 6.25 %	1/6 16.66666%	7/48 14.58333%
10 + 5 + 4	1/12 8.33333 %	1/16 6.25 %	1/12 8.33333 %	1/16 6.25 %	1/6 16.66666%	7/48 14.58333%

continuation: total value 19 (favorable values: 1, 2)

Variant	Favorable values				Total probability	
	1		2			
	A	B	A	B	C	D
9 + 9 + 1	1/16 6.25 %	1/24 4.16666 %	1/16 6.25 %	1/24 4.16666 %	7/48 14.58333%	1/8 12.5 %
9 + 8 + 2	1/12 8.33333 %	1/16 6.25 %	1/12 8.33333 %	1/16 6.25 %	7/48 14.58333%	1/8 12.5 %
9 + 7 + 3	1/12 8.33333 %	1/16 6.25 %	1/12 8.33333 %	1/16 6.25 %	1/6 16.66666%	7/48 14.58333%
9 + 6 + 4	1/12 8.33333 %	1/16 6.25 %	1/12 8.33333 %	1/16 6.25 %	1/6 16.66666%	7/48 14.58333%
9 + 5 + 5	1/12 8.33333 %	1/16 6.25 %	1/12 8.33333 %	1/16 6.25 %	1/6 16.66666%	7/48 14.58333%
8 + 8 + 3	1/12 8.33333 %	1/16 6.25 %	1/12 8.33333 %	1/16 6.25 %	1/6 16.66666%	7/48 14.58333%
8 + 7 + 4	1/12 8.33333 %	1/16 6.25 %	1/12 8.33333 %	1/16 6.25 %	1/6 16.66666%	7/48 14.58333%
8 + 6 + 5	1/12 8.33333 %	1/16 6.25 %	1/12 8.33333 %	1/16 6.25 %	1/6 16.66666%	7/48 14.58333%
7 + 7 + 5	1/12 8.33333 %	1/16 6.25 %	1/12 8.33333 %	1/16 6.25 %	1/6 16.66666%	7/48 14.58333%
7 + 6 + 6	1/12 8.33333 %	1/16 6.25 %	1/12 8.33333 %	1/16 6.25 %	1/6 16.66666%	7/48 14.58333%

total value 20 (favorable value 1)

Variant	A	B
11 + 9	3/49 6.12244 %	2/49 4.08163 %
10 + 10	4/49 8.16326 %	3/49 6.12244 %
11 + 8 + 1	1/24 4.16666 %	1/48 2.08333 %
11 + 7 + 2	1/16 6.25 %	1/24 4.16666 %
11 + 6 + 3	1/16 6.25 %	1/24 4.16666 %
11 + 5 + 4	1/16 6.25 %	1/24 4.16666 %
10 + 9 + 1	1/16 6.25 %	1/24 4.16666 %
10 + 8 + 2	1/12 8.33333 %	1/16 6.25 %
10 + 7 + 3	1/12 8.33333 %	1/16 6.25 %
10 + 6 + 4	1/12 8.33333 %	1/16 6.25 %
10 + 5 + 5	1/12 8.33333 %	1/16 6.25 %
9 + 9 + 2	1/12 8.33333 %	1/16 6.25 %
9 + 8 + 3	1/12 8.33333 %	1/16 6.25 %
9 + 7 + 4	1/12 8.33333 %	1/16 6.25 %
9 + 6 + 5	1/12 8.33333 %	1/16 6.25 %
8 + 8 + 4	1/12 8.33333 %	1/16 6.25 %
8 + 7 + 5	1/12 8.33333 %	1/16 6.25 %
8 + 6 + 6	1/12 8.33333 %	1/16 6.25 %
7 + 7 + 6	1/12 8.33333 %	1/16 6.25 %

The tables cover all situations in the distribution of two or three cards having a total value of 15, 16, 17, 18, 19 or 20 points.

However, all possible distribution situations are in the thousands because they include variants of 4, 5, 6, 7 and even many cards (for example, a hand with a total value 15 is made up of 9 cards: 2 + 3 + 2 + 2 + 2 + A + A + A + A), so they cannot be captured entirely in the tables.

To calculate the probabilities of obtaining one or more favorable values at a certain moment, the previously established formula remains fundamental:

– for a 1-deck game: $P = \dfrac{4 - nx}{52 - nv}$, if $x \neq 10$ and

$P = \dfrac{16 - nx}{52 - nv}$, if $x = 10$;

– for a 2-deck game: $P = \dfrac{8 - nx}{104 - nv}$, if $x \neq 10$ and

$P = \dfrac{32 - nx}{104 - nv}$, if $x = 10$;

– generally, for an *m*-deck game: $P = \dfrac{4m - nx}{52m - nv}$, if $x \neq 10$ and

$P = \dfrac{16m - nx}{52m - nv}$, if $x = 10$,

where *nx* is the number of cards showing having a value *x*, and *nv* the total number of cards showing.

This formula can be applied during a game through regular training in mental calculus.

In addition, presenting all probability results that correspond to all distribution situations is not even necessary because, as we said earlier, the vast majority of probabilities are low.

A strategy based on probability—that is: *I ask for an additional card if the probability of receiving a favorable value is at least ...%*—is indicated in blackjack (as well as in baccarat), because this game also includes confronting an adversary (the dealer) directly.

The real probability of finally winning is increased (when compared with the chosen threshold for decision) if we also take into account the hypothesis of the adversary's (the dealer) failure (unlike roulette, for example, which has no adversary to confront and the probabilities are calculated unilaterally).

Texas Hold'em Poker

This variation of poker is played at classical tables as well as online and uses a 52-card deck.

In this game, players are dealt two cards, face down. They are called the *hole* or *pocket cards*.

Five cards are then dealt in the middle of the table, face up. These cards are called the *community cards* and are shared by all the players at the table.

Players then try to construct the best possible five-card hand they can from the combination of community cards and their hole cards.

The player with the best hand wins.

The game is played as follows:

One player is the dealer. The two players to the left of the dealer make *blind* bets.

The first player makes a bet that is equal to half of the minimum bet at the table. This is known as the *small blind*. The second player makes a bet equal to the minimum table bet. This is the *big blind*.

Two cards are dealt face down to each player at the table.

The player to the left of the Big Blind begins play. He or she can choose to call (or match the bet on the table), raise or fold.

This play continues around to the left until the betting round is completed.

The first three face-up cards are then dealt at the same time. These three cards are called the *flop*.

The next betting round begins with the player to the left of the dealer and continues around the table to the left.

If there is no bet on the table, a player can bet or check (which means to neither bet nor fold, and the turn simply passes to the next player).

If there is a bet on the table, a player can call, raise or fold.

At the close of the betting round, the next card is dealt, face up. This card is called the *turn*.

The next round of betting begins. For this and the final round, the value of a bet increases to the table maximum.

After the close of the betting round, the final card is dealt. This card is known as the *river*.

The final betting round takes place. When all bets have been placed, the best hand takes the pot.

Here is a list of Texas Hold'em Poker hand rankings from strongest to weakest:

Royal Straight Flush – This hand is made of up of five cards of the same suit, 10 through A. It is a straight and a flush; for example: A♦ K♦ Q♦ J♦ 10♦;

Straight Flush – This hand is made up of five cards of the same suit ranked in succession; for example: Q♣ 10♣ 9♣ 8♣ 7♣;

Four-of-a-Kind (or quads) – A four-of-a-kind occurs when a player has all the cards of same value; for example: J♠ J♣ J♥ J♦;

Full House – A full house happens when a player has both three-of-a-kind and a pair. To determine if one full house ranks higher than another, first look at the three cards of the same kind. The one made up of cards of higher value wins. If they are both the same, then the player must compare the two paired cards; for example: 8♣ 8♦ 8♥ 9♣ 9♥;

Flush – A flush occurs when a player has five cards of the same suit. If the cards are also in succession (as in a straight), then the hand is called a straight flush. To determine which flush wins if there are more than one in a game, the person with the highest card in hand wins; for example: K♥ 8♥ 4♥ 3♥ 2♥;

Straight – A straight occurs when a player has five cards in ranked succession. Note that an ace can play both as a high card or a low card; for example: A♥ K♦ Q♦ J♠ 10♥ or A♥ 2♣ 3♥ 4♥ 5♣ or 7♠ 6♦ 5♣ 4♥ 3♣;

Three-of-a-Kind (trip or set) – A three-of-a-kind occurs when a player has exactly three cards of same value. There are two names for three-of-a-kind hands, depending on whether a player has two of them on the board or a pair in hand. They both rank the same. If a player has a pocket pair and hits one on the flop, then it is called a *set*. If the player has two cards on the flop and one in hand, then it is called three-of-a-kind or a trip; for example: 5♥ 5♦ 5♣ A♦ 3♥;

Two Pair – Two pair is when a player's best five cards create a pair twice. When comparing two hands, each with two pairs, the one with higher value always wins; for example: K♠ K♥ 8♣ 8♠ 7♦;

One Pair – One pair is when a player has two cards of the same value. When two persons have the same pair, the highest unpaired card comes into play. Remember, Texas Hold'em uses the best five cards, so the following is true: 9♥ 9♠ J♣ 5♥ 3♠ loses to 9♦9♣Q♦8♠7♥ and A♥ A♣ J♣ 6♦ 4♦ loses to A♦ A♠ J♦ 6♣ 5♦;

High Card – The high card hand is when a player has no pair. In the next example, the high card is *Q*. So the player's hand would be a *queen high*. That player would beat someone who had a *jack high* or any cards of lesser value, but even a pair of deuces would beat the queen high hand having no pair; for example: Q♥ 10♣ 5♣ 3♦ 2♥.

As in every card game, we deal here with a finite probability field, in which the events to be measured are occurrences of certain card combinations (three-card combinations, two-card combinations or one card) on the board, in your own hand and in your opponents' hands (called *favorable combinations)*.

The probability of achieving a specific final card formation is calculated by counting the favorable combinations that can occur.

When calculating the probabilities, the information taken into account at a specific moment in the game is given by parameters such as the number of certain cards from the community cards, the number of certain cards from the cards showing (community cards dealt plus cards from your own hand) and the number of opponents.

All calculations ignore any circumstantial information like viewing opponents' cards, removing certain cards from the deck, a particular distribution of cards in the deck before dealing or fraudulent dealing.

Among all games of chance, Texas Hold'em Poker is highly predisposed to probability-based decisions because, by the structure and dynamics of the game, the odds of an expected event change significantly from one gaming moment to another.

We have here the so-called *long-shot* odds (probabilities of events that are chronologically preceded by others, calculated by using information available at the moment before the first event), which consist of:

– **Own hand probabilities** – dealing with odds for you to achieve certain final card formations, calculated at two separate moments of the game: the three community cards dealt on the table (after the flop) and the four community cards dealt on the table (after the turn);

– **Opponents' hands probabilities** – dealing with odds for at least one opponent to achieve a certain final card formation, calculated in three separate moments of the game: three community cards dealt (after the flop), four community cards dealt (after the turn) and all five community cards dealt (after the river).

We also have the *immediate odds* (probabilities of events that follow immediately, calculated by using information of the moment just before that event): preflop odds, turn odds, odds of improving expected formations, and so on.

Although the long-shot odds are the most important when making a probability-based betting decision, because they give information about achieving specific final card formations (which is the technical goal of the game), the immediate odds could help in evaluating and creating an advantage during each betting dialogue.

Texas Hold'em Poker consists of two parts: the human action, namely, the betting dialogue and the random card distribution that results in building valuable formations.

Most players take into account the immediate odds before each betting moment, even though this information is not always too relevant for the final event, but they do so because they know their opponents refer to them, too.

Therefore, by making betting decisions based on immediate odds you can create an immediate advantage for yourself (for example, you can make other opponents fold).

This is the psychological element any poker game holds.

In fact, if the game consisted only of card distribution and building valuable formations, it would not be so spectacular.

We now present the most important immediate odds players refer to during the betting dialogue. The immediate odds categories are:

– preflop odds (odds of being dealt certain 2-card combinations as pocket cards before the flop)
– flop odds (odds for certain cards or certain 2- or 3-card combinations to be contained in the flop, calculated after first card distribution and right before the flop)
– turn odds (odds of a certain card being dealt as the turn card, calculated right after the flop)
– river odds (odds of a certain card being dealt as the river card, calculated right after the turn).

Preflop odds

Odds of being dealt AA

Favorable combinations: $C_4^2 = 6$.

Possible combinations: $C_{52}^2 = 1326$.

The probability is $P = C_4^2 / C_{52}^2 = 6/1326 = 0.452\%$ (220:1 as odds).

The same probability holds for any other pair.

Odds of being dealt AA or KK

These two events are incompatible, so their odds are added. Each has the probability C_4^2 / C_{52}^2, so

$P = 2C_4^2 / C_{52}^2 = 12/1326 = 0.904\%$ (110:1).

The same probability holds for any other two pairs.

Odds of being dealt KK, QQ or JJ

Similarly, we have $P = 3C_4^2 / C_{52}^2 = 18/1326 = 1.357\%$ (73:1).

The same probability stands for any other three pairs.

Odds of being dealt a pair

We have thirteen pair types and for each type the probability is C_4^2 / C_{52}^2. Then $P = 13C_4^2 / C_{52}^2 = 78/1326 = 5.882\%$ (16:1).

Odds of being dealt *AK* suited

For a specific symbol, we have one single (AK) suited combination. We have four symbols, thus four favorable combinations.

$P = 4/C_{52}^2 = 4/1326 = 0.301\%$ (330:1).

Odds of being dealt *AK* off suit

We have 4 x 4 = 16 combinations (AK). By subtracting the four suited, we find twelve favorable combinations.

$P = 12/C_{52}^2 = 12/1326 = 0.904\%$ (110:1).

Odds of being dealt *AQ* or *AJ* suited

These two events are incompatible, so we can add their odds.

For *AQ* suited, we have four favorable combinations and also four for *AJ*, so we have a total of eight favorable combinations.

$P = 8/C_{52}^2 = 8/1326 = 0.603\%$ (165:1).

Odds of being dealt *AQ* or *AJ* off suit

For *AQ* off suit we have twelve favorable combinations and also twelve for *AJ* off suit, so we have a total of twenty-four favorable combinations.

$P = 24/C_{52}^2 = 24/1326 = 1.809\%$ (54:1).

Odds of being dealt *KQ* suited

As in the *AK* suited case, $P = 4/C_{52}^2 = 4/1326 = 0.301\%$ (330:1).

Odds of being dealt *KQ* off suit

As in the *AK* off suit case,

$P = 12/C_{52}^2 = 12/1326 = 0.904\%$ (110:1).

Odds of being dealt *A* and < *J* suited

Favorable combinations: (A2), (A3), (A4), (A5), (A6), (A7), (A8), (A9), (A 10) suited.

Each of the nine types has four combinations, so we have a total of thirty-six favorable combinations.

$P = 36/C_{52}^2 = 36/1326 = 2.714\%$ (36:1).

Odds of being dealt A and a card lower than J off suit

Each of the combination types enumerated in the application above has twelve combinations in this case, so in total we have $12 \times 9 = 108$ favorable combinations.

$P = 108/C_{52}^2 = 108/1326 = 8.144\%$ (11:1).

Odds of being dealt two suited cards

For a specific symbol, we have $C_{13}^2 = 78$ favorable combinations. Multiplying by four (symbols), we find 312 favorable combinations.

$P = 4C_{13}^2/C_{52}^2 = 312/1326 = 23.529\%$ (3:1).

Odds of being dealt A or a pair

The two events whose probability we are looking for are:

(Ax) dealt, x *any card* and *(yy) dealt,* y *any card.*

These events are not incompatible (there are common favorable combinations, namely (AA)).

Therefore, the probability of being dealt A or pair is

$P = P((Ax)$ *dealt,* x *any card*$) + P((yy)$ *dealt,* y *any card*$) -$
$- P((AA)$ *dealt*$)$. (#)

For (Ax):

(Ax), with x different from A: $4 \times (52 - 4) = 192$ combinations

(AA): $C_4^2 = 6$ combinations

In total, we have 198 favorable combinations for (Ax).

For (yy): $13C_4^2 = 78$ favorable combinations

By replacing all these in formula (#), we find:

$P = 198/1326 + 78/1326 - 6/1326 = 270/1326 = 20.361\%$ (4:1).

Odds of being dealt any two cards higher than J

Favorable combinations: (xy), with x, y from the set (Q, K, A).

(QK): 16
(KA): 16
(QA): 16

In total, we have forty-eight favorable combinations.

$P = 48/C_{52}^2 = 48/1326 = 3.619\%$ (27:1).

Odds of being dealt at least one *A*

Favorable combinations: (Ax), in number of 198 (see *Odds of being dealt* A *or pair*).

$$P = 198 / C_{52}^2 = 198 / 1326 = 14.932\% \quad (6:1).$$

Odds of being dealt an *A* high

Favorable combinations: (Ax), with x different from A, numbering 192.

$$P = 192 / C_{52}^2 = 192 / 1326 = 14.479\% \quad (6:1).$$

Odds of being dealt suited connectors (23 to *KQ*)

We have eleven such suited connectors: (23), (34), (45), (56), (67), (78), (89), (9 10), (10, J), (JQ), (QK).

Each type has four combinations, so we have a total of $4 \times 11 = 44$ favorable combinations.

$$P = 44 / C_{52}^2 = 44 / 1326 = 3.318\% \quad (29:1).$$

Odds of being dealt connectors off suit

Each of the combination types described above has twelve combinations in this case, so we have a total of $12 \times 11 = 132$ favorable combinations.

$$P = 132 / C_{52}^2 = 132 / 1326 = 9.954\% \quad (9:1).$$

Odds of being dealt two adjacent and suited cards

We have twelve such combination types: (23), (34), (45), (56), (67), (78), (89), (9 10), (10, J), (JQ), (QK), (KA) and each has four combinations, so we have a total of $4 \times 12 = 48$ favorable combinations.

$$P = 48 / C_{52}^2 = 48 / 1326 = 3.619\% \quad (27:1).$$

Odds of being dealt two adjacent cards off suit

Each of the combination types previously described has twelve combinations in this case, so we have a total of $12 \times 12 = 144$ favorable combinations.

$$P = 144 / C_{52}^2 = 144 / 1326 = 10.859\% \quad (8:1).$$

Observation: These probabilities also hold for any of your opponents' hands, from outside view (that is, as long as we do not take into account your own pocket cards).

Pair against higher pair

Assuming you are dealt a pair (except *AA*), we can calculate the probability for at least one opponent being dealt a pair higher than yours.

We can find a unique formula that comprises all possible situations. Such a formula holds as variables:

n = number of your opponents

p = number of pairs (as a value type) higher than the one you hold.

For example, if you hold 88, there are six pairs higher than yours (99, 10 10, JJ, *QQ, KK, AA*), so p = 6; if you hold *KK*, there is only one pair higher (*AA*) and p = 1.

$0 \leq n \leq 22$, $1 \leq p \leq 12$.

For one opponent (a fixed one), the favorable combinations are (CC), *C* being a pair card higher than yours.

The number of these combinations is $pC_4^2 = 6p$.

The probability of one opponent being dealt a pair higher than yours is $6p / C_{50}^2 = 6p/1225$.

Many opponents could simultaneously hold such a pair.

For two opponents, the favorable combinations are (CC)(DD), *C* and *D* being pair cards higher than yours.

The number of these combinations is $6p(6p - 1)$ and the probability of two opponents being dealt higher pairs is

$[6p(6p - 1)]/(C_{50}^2 C_{48}^2) = p(6p - 1)/230300$.

For three opponents, the favorable combinations are (CC)(DD)(EE), *C, D* and *E* being pair cards higher than yours.

The number of these combinations is $6p(6p - 1)(6p - 2)$ and the probability of three opponents being dealt higher pairs is

$[6p(6p - 1)(6p - 2)]/(C_{50}^2 C_{48}^2 C_{46}^2) = p(6p - 1)(6p - 2)/238360500$.

For four opponents, we similarly find that the probability is $p(6p - 1)(6p - 2)(6p - 3)/225489033000$.

The final formula (for the probability that at least one opponent holds a higher pair) is:

$P = 6np/1225 - C_n^2 p(6p - 1)/230300 + C_n^3 p(6p - 1)(6p -$

$- 2)/238360500 - C_n^4 p(6p - 1)(6p - 2)(6p - 3)/225489033000 + ...$
(not all terms are shown) .

The maximal value for the last term shown is reached for $n = 22$ and $p = 12$. By replacing these values, we find 0.133 as the maximal value for $C_n^4 p(6p - 1)(6p - 2)(6p - 3)/225489033000$.

The next terms (not shown) have even a lower absolute value.

We can do the approximation:

$P = 6np/1225 - C_n^2 p(6p - 1)/230300 + C_n^3 p(6p - 1)(6p -$

$- 2)/238360500 = 6np/1225 - n(n - 1)p(6p - 1)/460600 + n(n - 1)$
$(n - 2)p(6p - 1)(6p - 2)/1430163000.$ (##)

This approximation does not affect the real result with more than $\pm 8\%$ for high values of n and p and much less for lower values of these parameters (Texas Hold'em is usually played by a maximum of ten players, so the error range is much lower).

Example:
You are dealt 99 in a game with seven opponents. Let us find the probability of at least one opponent holding a higher pair.

We have $n = 7$, $p = 5$ (10, J, Q, K, A). By replacing these in the formula above we find
$P = 6np/1225 - n(n - 1)p(6p - 1)/460600 + n(n - 1)(n - 2)p(6p -$
$- 1)(6p - 2)/1430163000 = 0.17142 - 0.01322 + 0.00059 = 0.15879$
$= 15.879\%$.

Particularly, for $p = 1$, the formula (##) becomes:
$P = 6n /1225 - 5n(n - 1)/460600 + 20n(n - 1)(n - 2)/1430163000.$

This particular formula holds for the case in which you are dealt *KK*. It returns the following results (for up to nine opponents):

Probabilities that at least one opponent holds *AA*, in case you are dealt *KK*

Opponents (n)	1	2	3	4	5	6	7	8	9
P (%)	5.877	5.387	4.897	4.408	3.918	3.428	2.938	2.448	1.959

A table of values for the probability (P) of one opponent (a specific one, not at least one) holding a pair higher than yours follows:

Probabilities that one opponent holds a higher pair

Own pair	22	33	44	55	66	77	88	99	T T	JJ	QQ	KK
P (%)	5.877	5.387	4.897	4.408	3.918	3.428	2.938	2.448	1.959	1.469	0.979	0.489

Other preflop odds

Odds of not holding an ace

The probability of no player holding an ace is calculated here from the outside view.

The complementary event is A – *At least one player holds an ace* (*an ace* means *one or more aces*). Let us find the formula for P(A). The probability we are looking for is then 1 – P(A).

For one player (a specific one), let us find the probability of that player holding an ace:

Favorable combinations: (Ax), x different from A, numbering $4 \times (52 - 4)$ and (AA), numbering $C_4^2 = 6$.

In total, we have 198 favorable combinations from $C_{52}^2 = 1326$ possible. The probability is $198/1326 = 99/661$.

A maximum of four players could hold an ace simultaneously.

The probability that two players hold an ace simultaneously follows:

Favorable combinations:

(Ax)(Ay), numbering $4 \times 48 \times 3 \times (52 - 3 - 2) = 27072$

(AA)(Ay), numbering $C_4^2 \times 2 \times (52 - 3 - 1) = 576$

(Ax)(AA), numbering 576

(AA)(AA), numbering $C_4^2 \times 1 = 6$, (x, y different from A).

In total, we have 28230 favorable combinations from $C_{52}^2 C_{50}^2$ possible. The probability is $941/54145$.

The probability that three players hold an ace simultaneously follows:

Favorable combinations:

(Ax)(Ay)(Az), numbering $4 \times 48 \times 3 \times 47 \times 2 \times 46$

(AA)(Ay)(Az), numbering $6 \times 2 \times 48 \times 1 \times 47$

(Ax)(AA)(Az), numbering $6 \times 2 \times 48 \times 1 \times 47$

(Ax)(Ay)(AA), numbering $6 \times 2 \times 48 \times 1 \times 47$,

(x, y, z different from A).

In total, we have $12 \times 48 \times 47 \times 95$ favorable combinations from $C_{52}^2 C_{50}^2 C_{48}^2$ possible. The probability is $228/162435$.

The probability that four players hold an ace simultaneously follows:

Favorable combinations: (Ax)(Ay)(Az)(At), numbering

$4 \times 48 \times 3 \times 47 \times 2 \times 46 \times 1 \times 45$ (x, y, z, t different from A).

Possible combinations: $C_{52}^2 C_{50}^2 C_{48}^2 C_{46}^2$.

The probability is $16/270725$.

Now applying the inclusion-exclusion principle, we find:

$P(A) = 99n/661 - 941\,C_n^2/54145 + 228\,C_n^3/162435 - 16\,C_n^4/270725$

$= 99n/661 - n(n-1)941/108290 + 38n(n-1)(n-2)/162435 -$

$- 2n(n-1)(n-2)(n-3)/812175$, where n is the number of players.

Therefore, the probability we are looking for is:

$1 - 99n/661 + 941n(n-1)/108290 - 38n(n-1)(n-2)/162435 +$

$+ 2n(n-1)(n-2)(n-3)/812175$.

By giving values to n, we find:

For n = 2, the probability is 71.783%

For n = 3, the probability is 60.141%

For n = 4, the probability is 49.962%

For n = 5, the probability is 41.118%

For n = 6, the probability is 33.486%

For n = 7, the probability is 26.949%

For n = 8, the probability is 21.396%

For n = 9, the probability is 16.723%

For n = 10, the probability is 12.831%.

You have one ace, probability of no one else having an ace

We have 50 unknown cards. The complementary event is

A – *At least one player holds an ace.*

The probability we are looking for is then $1 - P(A)$.

For one player (a specific one), let us find the probability of that player holding an ace:

Favorable combinations:

(Ax), x different from A, in number of 3 x (50 – 3) = 141

(AA), numbering $C_3^2 = 3$.

In total, we have 144 favorable combinations from $C_{50}^2 = 1225$ possible. The probability is 144/1225 .

A maximum of three players could hold an ace simultaneously.

The probability that two players hold an ace simultaneously follows:

Favorable combinations:

(Ax)(Ay), numbering 3 x 47 x 2 x 46

(AA)(Ay), numbering C_3^2 x 1 x 47

(Ax)(AA), numbering C_3^2 x 1 x 47,

(x, y different from A).

In total, we have 13254 favorable combinations from $C_{50}^2 C_{48}^2$ possible. The probability is 47/4900.

The probability that three players hold an ace simultaneously follows:

Favorable combinations are (Ax)(Ay)(Az), numbering

3 x 47 x 2 x 46 x 1 x 45 (x, y, z different from A), from $C_{50}^2 C_{48}^2 C_{46}^2$ possible. The probability is 1/2450.

Now applying the inclusion-exclusion principle, we find:

$P(A) = 144n/1225 - 47 C_n^2 /4900 + C_n^3 /2450 = 144n/1225 -$

$- 47n(n - 1)/9800 + n(n - 1)(n - 2)/14700$,

where n is the number of your opponents.

Therefore, the probability we are looking for is:

$1 - 144n/1225 - 47n(n - 1)/9800 + n(n - 1)(n - 2)/14700$.

By giving values to n, we find:

For n = 1 (2 players), the probability is 88.244%

For n = 2 (3 players), the probability is 77.448%

For n = 3 (4 players), the probability is 67.571%

For n = 4 (5 players), the probability is 58.571%

For n = 5 (6 players), the probability is 50.408%
For n = 6 (7 players), the probability is 43.040%
For n = 7 (8 players), the probability is 36.428%
For n = 8 (9 players), the probability is 30.530%
For n = 9 (10 players), the probability is 25.306%.

One ace on the flop, odds that someone else has one ace

This calculation is made from the outside view (only flop cards are seen).

We have 49 unknown cards. The event is A – *At least one player has one ace.*

The probability that one player (a specific one) holds one ace:

Favorable combinations: (Ax), x different from A, numbering $3 \times (49 - 3) = 138$. Possible combinations: $C_{49}^2 = 1176$.

The probability is $138/1176 = 23/196$.

A maximum of three players could hold one ace simultaneously.

The probability that two players hold one ace simultaneously follows:

Favorable combinations: (Ax)(Ay), x, y different from A, numbering $3 \times 46 \times 2 \times 45$. Possible combinations: $C_{49}^2 C_{47}^2$.

The probability is $3 \times 46 \times 2 \times 45/(C_{49}^2 C_{47}^2) = 45/4606$.

The probability that three players hold one ace simultaneously follows:

Favorable combinations: (Ax)(Ay)(Az), x, y, z different from A, numbering $3 \times 46 \times 2 \times 45 \times 1 \times 44$.

Possible combinations: $C_{49}^2 C_{47}^2 C_{45}^2$.

The probability is $3 \times 46 \times 2 \times 45 \times 1 \times 44/(C_{49}^2 C_{47}^2 C_{45}^2) = 1/2303$.

Now applying the inclusion-exclusion principle, we find:

$P(A) = 23n/196 - 45 C_n^2 /4606 + C_n^3 /2303 = 23n/196 - 45n(n - 1) /9212 + n(n - 1)(n - 2)/13818$,

where n is the number of players.

By giving some values to n, we find:
For n = 5, the probability is 49.337%
For n = 4, the probability is 41.250%
For n = 3, the probability is 32.316%.

Flop odds

The moment for calculating flop odds is after first card distribution and before the flop.

We want to find the probabilities of the flop improving an expected formation (three-of-a-kind, four-of-a-kind, flush or straight).

We have two cards showing (your own pocket cards), 50 unknown cards and $C_{50}^3 = 19600$ possible combinations for the flop cards.

Odds of improving three-of-a-kind or four-of-a-kind

Let C be a card from an expected formation and c the number of your own pocket C-cards (as value). c could be 0, 1 or 2.

The outs (C-cards from the 50 unknown) number $4 - c$.

1) Odds of the flop containing exactly one C-card
The favorable combinations are (Cxy), with x and y different from C, numbering $(4 - c)\,C_{46+c}^2 = (4 - c)(45 + c)(46 + c)/2$.

The probability is

$P = (4 - c)\,C_{46+c}^2 /19600 = (4 - c)(45 + c)(46 + c)/39200$.

2) Odds of the flop containing exactly two C-cards
The favorable combinations are (CCx), with x different from C, numbering $C_{4-c}^2 (46 + c) = (3 - c)(4 - c)(46 + c)/2$.

The probability is $P = (3 - c)(4 - c)(46 + c)/39200$.

3) Odds of the flop containing three C-cards
If $c = 2$, then $P = 0$

If $c = 1$ or $c = 0$, we have $C_{4-c}^3 = (2 - c)(3 - c)(4 - c)/6$ favorable combinations and $P = (2 - c)(3 - c)(4 - c)/117600$.

These formulas generate the following table of values:

c	One expected card	Two expected cards	Three expected cards
0	21.122%	1.408%	0.020%
1	16.545%	0.719%	0.005%
2	11.510%	0.244%	0

Odds of improving a flush

Let S be a symbol and s the number of your own pocket cards holding this symbol (S-cards); s could be 0, 1 or 2.

The number of outs is $13 - s$.

1) Odds of the flop containing exactly one S-card

The favorable combinations are (Sxy), with x and y different from S, numbering $(13 - s)\,C^2_{37+s} = (13 - s)(36 + s)(37 + s)/2$.

The probability is $P = (13 - s)(36 + s)(37 + s)/39200$.

2) Odds of the flop containing exactly two S–cards

The favorable combinations are (SSx), with x different from S, numbering $C^2_{13-s}(37 + s) = (12 - s)(13 - s)(37 + s)/2$.

The probability is $P = (12 - s)(13 - s)(37 + s)/39200$.

3) Odds of the flop containing three S-cards

The favorable combinations are (SSS), numbering $C^2_{13-s} = (11 - s)(12 - s)(13 - s)/6$.

The probability is $P = (11 - s)(12 - s)(13 - s)/117600$.

These formulas generate the following table of values:

366

Odds of improving a flush on the flop

s	One expected card	Two expected cards	Three expected cards
0	44.173%	14.724%	1.459%
1	43.040%	12.795%	1.122%
2	41.586%	10.943%	0.841%

Odds of improving a straight

Let m be the number of your own pocket cards that are different (as value) and belong to the expected straight (for example, if 34567 is the expected straight and you hold 35, then m = 2; for the same straight, if you hold 77, then m = 1); m could be 0, 1 or 2.

1) Odds of the flop containing exactly one card from an expected straight

The favorable combinations are (Cxy), where C is a card from an expected straight, different from your pocket cards and x, y are different from the expected cards left (x, y could be even C).

The number of these combinations is

$4(5 - m) C_{33+4m}^2 = 2(5 - m)(32 + 4m)(33 + 4m).$

The probability is $P = (5 - m)(32 + 4m)(33 + 4m)/9800.$

2) Odds of the flop containing exactly two cards from an expected straight

The favorable combinations are (CDx), where C, D are cards from an expected straight, different from your pocket cards and x is different from the expected cards left (x could be even C or D).

The number of these combinations is

$16 C_{5-m}^2 (36 + 4m) = 8(4 - m)(5 - m)(36 + 4m).$

The probability is $P = (4 - m)(5 - m)(36 + 4m)/2450.$

3) Odds of the flop containing three cards from an expected straight

The favorable combinations are (CDE), where C, D and E are cards from an expected straight, different from your pocket cards.

The number of these combinations is

$64\, C_{5-m}^3 = 32(3-m)(4-m)(5-m)/3$ and the probability is

$P = 2(3-m)(4-m)(5-m)/3675$.

These formulas generate the following table of values:

Odds of improving a straight on the flop

m	One expected card	Two expected cards	Three expected cards
0	53.877%	29.387%	3.265%
1	54.367%	19.591%	1.306%
2	50.204%	10.775%	0.326%

Example:
You hold JK. Let us find the probability of improving an expected (9 10 JQK) straight on the flop.

We have m = 2. By looking at the table, we find 50.204% for one card, 10.775% for two cards and 0.326% for three cards.

The same odds hold for a (10 JQKA) straight.

Flop odds when you hold specific cards

When you hold AK suited

Let S be the symbol of your suit.

Q, J, 10 (royal flush)
Favorable combinations: 1 (Q 10 J).
Possible combinations: $C_{50}^3 = 19600$.
Probability: $1/19600 = 0.0051\%$ (19599:1).

Improves to four-of-a-kind
Favorable combinations: (AAA) – 1, (KKK) – 1.
In total, we have two favorable combinations from 19600
possible. Probability: $2/19600 = 1/9800 = 0.0102\%$ (9799:1).

Improves to full house
Favorable combinations:

(AAK) – numbering $C_3^2 \times 3 = 9$

(KKA) – numbering $C_3^2 \times 3 = 9$

In total, we have eighteen favorable combinations.
Probability: $18/19600 = 1/1089 = 0.0918\%$ (1088:1).

Flopped flush
Favorable combinations: (SSS), without (10 J Q of symbol S),
numbering $C_{11}^3 - 1 = 164$.
Probability: $164/19600 = 0.836\%$ (119:1).

Pair of A or K with four flush
Favorable combinations:

(ASS), numbering $3 \times C_{11}^2 = 165$

(KSS), numbering $3 \times C_{11}^2 = 165$

In total, we have 330 favorable combinations.
Probability: $330/19600 = 1.683\%$ (58:1).

Two of your own suit
Favorable combinations: (SSx), with x different from S,
numbering $C_{11}^2 (50 - 11) = 2145$.
Probability: $2145/19600 = 10.943\%$ (8:1).

Any Q, J, 10
Favorable combinations: (QJ 10), numbering $4 \times 4 \times 4 = 64$.
Probability: $64/19600 = 0.326\%$ (305:1).

QQ or less, one or less of your suit
The favorable combinations are:
(xxS), x different from S (as symbol), x different from A and K
(as value) – numbering $11 \times C_3^2 \times 11$ and
(xxy), y different from S (as symbol), x different from A and K
(as value) – numbering $11 \times C_4^2 \times (50 - 11)$.
We now must subtract the common combinations, which are the
trips (xxx), none having the symbol S. These number $11 \times (C_4^3 - 1)$.
So, the exact number of favorable combinations is
$11 \times C_3^2 \times 11 + 11 \times C_4^2 \times (50 - 11) - 11 \times (C_4^3 - 1) = 3267$.
Probability: $3267/19600 = 16.668\%$ (5:1).

Flop two pairs
Favorable combinations:
(AKx), x different from A and K, numbering $3 \times 3 \times (50 - 3 - 3)$
(Axx), x different from A and K, numbering $3 \times 11 \times C_4^2$
(Kxx), x different from A and K, numbering $3 \times 11 \times C_4^2$.
In total, we have 892 favorable combinations.
Probability: $892/19600 = 4.551\%$ (21:1).

222 to QQQ
Favorable combinations: (222), ..., (QQQ), numbering
$11 \times C_4^3 = 44$.
Probability: $44/19600 = 0.224\%$ (444:1).

A or *K*
Favorable combinations:

(Axy), numbering $3 \times C_{44}^2$

(Kxy), numbering $3 \times C_{44}^2$

(AAx), numbering 3×4

(AAA), in number of 1

(KKx), numbering 3×44

(KKK), numbering 1

(AKx), numbering 9×44

(AAK), numbering $C_3^2 \times 3$

(KKA), numbering $C_3^2 \times 3$

(x, y different from *A* and *K*).
In total, we have 6356 favorable combinations.
Probability: $6356/19600 = 32.428\%$ (2:1).

AA or *KK* making trips
Favorable combinations:

(AAx), x different from *A* and *K*, numbering $C_3^2 \times 44$

(KKx), x different from *A* and *K*, numbering $C_3^2 \times 44$.

In total, we have 264 favorable combinations and the probability is $264/19600 = 1.346\%$ (73:1).

When you hold *KK*

Flopping four-of-a-kind
Favorable combinations: (KKx), x different from *K*, numbering $1 \times (50 - 2) = 48$.
Probability: $48/C_{50}^3 = 48/19600 = 0.244\%$ (407:1).

KAA
Favorable combinations: (KAA), numbering $2C_4^2 = 12$.
Probability: $12/19600 = 0.061\%$ (1632:1).

AK any card
Favorable combinations: (AKx), x different from A and K, numbering $4 \times 2 \times (50 - 4 - 2) = 352$.
Probability: $352/19600 = 1.795\%$ (55:1).

Making any full house
Favorable combinations:

(xxx), x different from K, numbering $12C_4^3 = 48$

(Kxx), x different from K, numbering $2 \times 12 \times C_4^2 = 144$.

In total, we have 192 favorable combinations for a full house.
Probability: $192/19600 = 0.979\%$ (101:1).

Trips
Favorable combinations: (Kxy), x, y different from K, x different from y – numbering $2C_{48}^2$ minus the number of (Kxx) combinations, that is, $2256 - 144 = 2112$ (see the paragraph on *making any full house*).
Probability: $2112/19600 = 10.775\%$ (8:1).

AAA
Favorable combinations: (AAA), numbering $C_4^3 = 4$.
Probability: $4/19600 = 1/4900 = 0.020\%$ (4899:1).

*AA*7
Favorable combinations: (AA7), numbering $C_4^2 \times 4 = 24$.
Probability: $24/19600 = 0.122\%$ (816:1).

Any pair with no *K* or *AA*
Favorable combinations: (xxy), x different from K and A, y different from K, x different from y – numbering $11 \times C_4^2 \times (50 - 2 - 2) - 11 \times C_4^3 = 2860$.
Probability: $2860/19600 = 14.591\%$ (6:1).

A and any other cards besides a *K*
Favorable combinations: (Axy), x, y different from A and K – numbering $4 \times C_{50-4-2}^2 = 4 C_{44}^2 = 3784$.
Probability: $3784/19600 = 19.306\%$ (4:1).

When you hold *QJ* off suit

Flop four *Q*'s or *J*'s
Favorable combinations: (QQQ) – 1, (JJJ) – 1.
We have two favorable combinations from $C_{50}^3 = 19600$ possible.
Probability: 2/19600 = 1/9800 = 0.010% (9799:1).

QQJ or JJQ
Favorable combinations: (QQJ) and (JJQ), numbering $C_3^2 \times 3 = 9$
each. In total, we have eighteen favorable combinations.
Probability: 18/19600 = 0.091% (1088:1).

AK 10
Favorable combinations: (AK 10), numbering 4 x 4 x 4 = 64.
Probability: 64/19600 = 0.326% (305:1).

K 10 9 or 89 10
Favorable combinations: (K 10 9) and (89 10), numbering
4 x 4 x 4 each. In total, we have 128 favorable combinations.
Probability: 128/19600 = 0.653% (152:1).

Flopping any straight
The possible straights are: (89 10 JQ), (9 10 JQK) and
(10 JQKA).
The favorable combinations for these straights are: (89 10),
(9 10 K) and (10 KA), and number 4 x 4 x 4 = 64 each.
In total, we have 192 favorable combinations.
Probability: 192/19600 = 0.979% (101:1).

Open-ended straight
The possible open-ended straights are: (9 10 JQ) and (10 JQK).
Favorable combinations:
(9 10 x), *x* different from 8 and *K*, numbering
4 x 4 x (50 – 2 – 4 – 4) = 640
(10 Kx), *x* different from 9 and *A*, numbering
4 x 4 x (50 – 2 – 4 – 4) = 640.
In total, we have 1280 favorable combinations.
Probability: 1280/19600 = 6.530% (14:1).

Three suited cards of a symbol you hold

Let S be one of the symbols you hold.

Favorable combinations: (SSS), numbering $C_{12}^3 = 220$.

In total, we have 440 favorable combinations (if you hold two different symbols).

Probability: $440/19600 = 2.244\%$ (44:1).

Flopping trips (no full house)

Favorable combinations:

(xxx), x different from J and Q, numbering $11 \times C_4^3 = 44$

(JJx), x different from J and Q, numbering $C_3^2 \times (50 - 3 - 3) = 3 \times 44$

(QQx), x different from J and Q, numbering $C_3^2 \times (50 - 3 - 3) = 3 \times 44$.

In total, we have 308 favorable combinations.

Probability: $308/19600 = 1.571\%$ (63:1).

No A or K

Favorable combinations: (xyz), x, y and z different from A and K.

Let us count the combinations holding A or K:

(Axy), numbering $4\,C_{50-4-4}^2 = 4\,C_{44}^2$

(AAx), numbering $C_4^2(50 - 4 - 4) = 6 \times 44$

(AAA), numbering $C_4^3 = 4$

(Kxy), numbering $4\,C_{50-4-4}^2 = 4\,C_{44}^2$

(KKx), numbering $C_4^2(50 - 4 - 4) = 6 \times 44$

(KKK), numbering $C_4^3 = 4$

(AKx), numbering $4 \times 4 \times (50 - 4 - 4) = 16 \times 44$

(AKK), numbering $4\,C_4^2 = 24$

(KAA), numbering $4\,C_4^2 = 24$

(x, y different from A and K).

In total, we have 8856 such combinations.

The number of favorable combinations is $C_{50}^3 - 8856 = 10744$.

Probability: $10744/19600 = 54.816\%$ (1:1).

Q's up or J's up

Favorable combinations:

(Jxx), x different from J, Q, K, A, numbering $3 \times 9 \times C_4^2 = 162$

(Qxx), x different from J, Q, K, A, numbering $3 \times 9 \times C_4^2 = 162$

(QJx), x different from J, Q, numbering $3 \times 3 \times (50 - 3 - 3) = 396$.

Among these combinations, there are no common combinations to subtract.

In total, we have 720 favorable combinations and the probability is $720/19600 = 3.673\%$ (26:1).

Other flop odds

No pair improving to a pair on the flop

Let us denote by (CD) the hole cards (C different from D as value).

Favorable combinations of the flop:

(Cxy), x, y different from C and D, x different from y – numbering $3(C_{50-3-3}^2 - 11\,C_4^2)$

(Dxy), x, y different from C and D, x different from y – numbering $3(C_{50-3-3}^2 - 11\,C_4^2)$

In total, we have 5280 favorable combinations from $C_{50}^3 = 19600$ possible.

Probability: $5280/19600 = 26.938\%$ (3:1).

Flop being all of a kind

– From the outside view (no pocket cards seen):

For each card (as a value), we have $C_4^3 = 4$ favorable combinations for this event.

In total, we have $13 \times 4 = 52$ favorable combinations, from $C_{52}^3 = 22100$ possible.

The probability is $52/C_{52}^3 = 52/22100 = 0.235\%$ (424:1).

– When taking into account your own pocket cards:

If you hold a pair, the probability of the flop being all of a kind is $12 \times 4/12 \times 4/C_{50}^3 = 48/19600 = 0.244\%$ (407:1).

If you hold unpaired cards, the probability of flop being all of a kind with the triple card different from those you hold is

$11 \times 4 / C_{50}^{3} = 44/19600 = 0.224\%$ (444:1).

Flop being unpaired

This problem was solved in the chapter titled *Probability Calculus Applications.*
The probability is 16104/22100 = 72.868%.

Turn odds

The moment for calculating turn odds is after flop and before the turn.
We want to find the probabilities for the turn card to complete an expected formation (three-of-a-kind, four-of-a-kind, flush or straight).
We have five seen cards (your own pocket cards and the flop cards) and 47 unknown cards.

Completing three-of-a-kind or four-of-a-kind

Let C be a card from an expected formation and c the number of seen C-cards (as a value).
c could be 2 or 3 (for 0 or 1, three-of-a-kind cannot be achieved and for 4, four-of-a-kind is already achieved).
The outs (the C-cards from the 47 unknown) number $4 - c$.
The probability for the turn card to be C is
$P = (4 - c)/47 = 4.255\%$ (22:1) for c = 2 or 2.127% (46:1) for c = 3.

Completing a flush

Let S be the expected symbol and s the number of seen S-cards.
The problem of completing a flush makes sense only for $s = 4$.
The probability that the turn card is S is
$P = (13 - s)/47 = 9/47 = 19.148\%$ (4:1).

Completing a straight

Let m be the number of seen different cards from an expected straight.
The problem of completing a straight makes sense only for $m = 4$.
The probability that the turn card completes the straight is
$P = 4(5 - m)/47 = 4/47 = 8.510\%$ (11:1).

Other odds

The results that follow are not immediate odds, according to the definition given earlier (they are associated with events preceded by others).

They are presented in this section because they are part of the odds taken into account in making betting decisions at various gaming moments.

QQ not having A or K hit by the river

Hypothesis: your own hand is dealt QQ. Event: A or K not hit by the river.

First, let us find the probability of A or K being hit by the river:
The favorable combinations for this event (the final board configuration) are:

(Axyzt), numbering $4C_{42}^4$

(AAxyz), numbering $6C_{42}^3$

(AAAxy), numbering $4C_{42}^4$

(AAAAx), numbering 1×42

(Kxyzt), numbering $4C_{42}^4$

(KKxyz), numbering $6C_{42}^3$

(KKKxy), numbering $4C_{42}^4$

(KKKKx), numbering 1×42

(AKxyz), numbering $16C_{42}^3$

(AKKxy), numbering $4 \times 6 \times C_{42}^2$

(AKKKx), numbering $4 \times 4 \times 42$

(AKKKK), numbering 4

(AAKxy), numbering $6 \times 4 \times C_{42}^2$

(AAKKx), numbering $6 \times 6 \times 42$

(AAKKK), numbering 6×4

(AAAKx), numbering $4 \times 4 \times 42$

(AAAKK), numbering 4×6

(AAAAK), numbering 4

(where x, y, z and are different from A and K).

In total, we have 1268092 favorable combinations for the event A *or K hit by river.*

The total number of possible combinations for the final board is $C_{50}^5 = 2118760$. The probability is $1268092 / 2118760 = 59.850\%$.

Then, the probability for the event *A or K not hit by the river* is $1 - (1268092 / 2118760) = 40.149\%$.

Four flush improving

Hypothesis: four card of the same symbol in your own hand plus the flop. Event: *at least one symbol of your suit hit by the river.*

Let S be the symbol of that suit.

The favorable 2-card combinations of the turn plus river cards are:

(Sx), with x different from S, numbering $9 \times (47 - 9)$ and

(SS), numbering $C_9^2 = 36$.

In total, we have 378 favorable combinations from $C_{47}^2 = 1081$ possible.

The probability is $378/1081 = 34.967\%$ (2:1).

Open-ended straight flush improving

Hypothesis: four consecutive cards from a straight flush in your own hand plus the flop. Event: *one or both of the two expected cards to be hit by the river.*

Let C and D be the cards needed (unique cards as a value and a symbol).

The favorable combinations of turn and river cards are:

(Cx), with x different from C and D, numbering $47 - 2$

(Dx), with x different from C and D, numbering $47 - 2$

(CD), numbering 1.

In total, we have ninety-one favorable combinations from $C_{47}^2 = 1081$ possible.

The probability is $91/1081 = 8.418\%$ (11:1).

Open-ended straight improving

Hypothesis: four consecutive cards from a straight in your own hand plus the flop. Event: *one or both of the two expected cards (as a value) to be hit by the river.*

Let C and D be the cards needed (as a value).

Favorable combinations of turn and river cards are:

(Cx), with x different from C and D, numbering $4 \times (47 - 8)$

(Dx), with x different from C and D, numbering $4 \times (47 - 8)$

(CD), numbering 16

(CC), numbering C_4^2

(DD), numbering C_4^2.

In total, we have 340 favorable combinations from $C_{47}^2 = 1081$ possible.

The probability is $340/1081 = 31.452\%$ (2:1).

Two pairs making a full house

Hypothesis: two pairs in your own hand plus the flop.

Event: *making a full house by the river.*

Let us denote by (TTDDC) the cards from own hand and the flop, with T, D and C mutually different (as a value).

Favorable combinations of turn plus river cards for achieving a full house:

(Tx), x different from T, D and C, numbering $2 \times (47 - 2 - 2 - 3)$

(Dx), x different from T, D and C, numbering $2 \times (47 - 2 - 2 - 3)$

(TT), numbering 1

(TD), numbering 2×2

(DD), numbering 1

(CC), numbering C_3^2

(TC), numbering 2×3

(DC), numbering 2×3.

In total, we have 181 favorable combinations from $C_{47}^2 = 1081$ possible. The probability is $181/1081 = 16.743\%$ (5:1).

379

Trips improving to full house or better

Hypothesis: three of the same cards (as a value) in your own hand plus the flop. Event: *making a full house or four-of-a-kind by the river.*

Let us denote by (TTTCD) the cards (*T*, *C* and *D* are mutually different as a value).

The favorable combinations of turn plus river cards are:

(Cx), *x* different from *T*, *C* and *D*, numbering $3 \times (47 - 1 - 3 - 3)$

(Dx), *x* different from *T*, *C* and *D*, numbering $3 \times (47 - 1 - 3 - 3)$

(CT), numbering 3×1

(DT), numbering 3×1

(DC), numbering 3×3

(xx), *x* different from *T*, *C* and *D*, numbering $10C_4^2$

(CC), numbering C_3^2

(DD), numbering C_3^2

(Tx), *x* different from *T*, *C* and *D*, numbering $47 - 1 - 3 - 3$.

In total, we have 361 favorable combinations from $C_{47}^2 = 1081$ possible.

The probability is $361/1081 = 33.395\%$ (2:1).

One pair on the flop improving by the river

Hypothesis: hole and flop cards: (CCxyz), x, y, z different from *C* (as a value).

Favorable combinations of turn and river cards:

(Cx), *x* different from *C*, numbering $2 \times (47 - 2) = 90$

(CC), numbering 3

(xx), *x* different from *C*, numbering $12C_4^2 = 72$.

In total, we have 165 favorable combinations from $C_{47}^2 = 1081$ possible.

The probability is $165/1081 = 15.263\%$ (6:1).

Pocket pair improving to trips after the flop

Pocket cards: (CC) Flop: (xyz), x, y, z different from *C*

Favorable combinations of turn and river cards: (Cx), *x* different from *C*, numbering $2 \times (47 - 2) = 90$, from $C_{47}^2 = 1081$ possible.

The probability is $90/1081 = 8.325\%$ (11:1).

Backdoor flush

Hypothesis: hole and flop cards: (SSSxy), x, y different from S (as a symbol).

Favorable combinations of turn and river cards: (SS), numbering $C_{10}^2 = 45$, from 1081 possible.

The probability is $45/1081 = 4.162\%$ (23:1).

We presented here a large collection of immediate odds.

As we said earlier, the long-shot odds (probabilities of events dealing with 1, 2 or 5 cards to come) are the most important in making decisions.

The long-shot odds for your own hand and also for your opponents' hands are calculated at different stages of the game.

Overall, they are numerous (they fill dozens of tables).

Anyone interested can find all these odds and all the math behind this game in my book, *Texas Hold'em Odds,* along with the algorithms of calculus and examples of how to use the results.

As an example, we present here the flush odds for opponents (only the results, without the calculus algorithm).

Opponents' hands probabilities – Flush odds

Let us denote by C the event: *At least one opponent makes a (SSSSS) flush formation by using five cards from his or her own hand and the community cards, no matter how many from each.*

We calculate the probability $P(C)$ in the gaming moment when only three community cards are dealt (after the flop and before the turn).

The variables the probability formula depends on are:

s'' = number of community S-cards (cards with symbol S)

s = number of seen S-cards (from own hand and board)

n = number of opponents

These variables must fit the conditions: s'', s, n are natural numbers, $s'' \leq 3$, $s \leq 5$, $0 \leq s - s'' \leq 2$, $1 \leq n \leq 22$.

381

For a (SSSSS) flush formation to be finally achieved, a minimal initial condition is that the community cards (the three dealt) must contain a minimum one of the S-cards ($s'' \geq 1$).

Table of values of probability that at least one opponent will make a (SSSSS) flush formation by river

$s''=0$	$s''=1$			$s''=2$			$s''=3$		
	1	2	3	2	3	4	3	4	5
$\dfrac{s}{n}$									
1 0	0.277%	0.184%	0.117%	3.515%	2.606%	1.860%	19.409%	16.048%	12.895%
2 0	0.546%	0.365%	0.233%	6.359%	4.777%	3.454%	30.402%	23.533%	20.998%
3 0	0.806%	0.542%	0.294%	8.595%	6.894%	4.439%	37.557%	32.861%	26.517%
4 0	1.059%	0.714%	0.459%	10.373%	8.053%	6.890%	41.889%	37.139%	32.376%
5 0	1.303%	0.882%	0.569%	12.041%	9.350%	7.057%	45.721%	40.553%	35.631%
6 0	1.539%	1.046%	0.677%	13.610%	10.579%	7.945%	49.129%	43.599%	37.987%
7 0	1.767%	1.206%	0.783%	15.081%	11.738%	8.819%	52.117%	46.276%	40.314%
8 0	1.987%	1.362%	0.887%	16.455%	12.871%	9.678%	54.685%	48.903%	42.610%
9 0	2.200%	1.493%	0.990%	17.832%	13.794%	10.524%	57.514%	50.894%	44.874%

THE PROBABILITY-BASED STRATEGY

As we saw in the section dedicated to the psychology of probability, the mathematical probability is frequently chosen as a decision criterion in daily life.

The motivation for such a choice generally has a psychological, therefore, subjective nature, and any denying of it may find solid arguments among the relativities of probability presented in the section dedicated to this subject.

But there is a particular type of decisional behavior in which choosing probability as unique criterion of decision can be mathematically justified. It is about using the probability-based strategy.

This kind of strategy is generally suitable for games, but it can be extended by analogy to certain situations in daily life as well.

The term *strategy* assumes, directly or indirectly, a certain repetitive action, even though that strategy is applied in a singular context.

When a player acts in a certain way in a gaming situation as a result of a previously adopted strategy, this action is repeated the next time a similar situation is encountered (because that strategy is considered *good* or *optimum* with respect to some player's objective or subjective criteria).

Thus, the strategy *I act in way* x *in situation* y, even if applied in an isolated situation, assumes intrinsically a collective, even infinite context: *I will act in way* x *in any hypothetical situation of type* y *from now on.*

If we refer to the probability-based strategy, the infinite collective is still much more present in this concept, because the mathematical definition of probability uses such a collective.

What does a probability-based strategy actually mean?

The following statement: *The strategy using criteria for evaluation and comparison of probabilities of expected events in making decisions for accomplishing the proposed goal* might be a general definition.

In games, the declared goal of a player in a certain gaming situation might be: winning cash immediately, obtaining a superior position within the game, avoiding a disadvantageous situation, and so on.

Of course, fun, entertainment or experiencing certain emotions that are specific to competition or risk might also be a player's goals.

In this work we ignore these possible additional goals, assuming the player aims only at a material winning or at improving a position.

As we said earlier, a probability-based strategy consists of decisions resulting from the evaluation and comparison of probability results.

These might be decisions involving choosing a specific playing variant at a certain moment, but also of choosing a certain game.

In this chapter we try to prove that a probability-based strategy is optimum (with respect to other types of strategies), at least at the theoretical level.

To build a mathematical model that represents this problem as accurately as possible, we must rigorously define *optimum strategy*, within a set of strategies standing for the basis of choice.

To do this, we must bear in mind that:

– The goal of choosing a strategy is to gain the advantage during a game at practical level (namely, a favorable event, which is the result of an experiment or a succession of experiments); thus, even if we look for a purely theoretical motivation, this motivation cannot ignore the practical results.

– As long as the probability-based strategy uses the concept of mathematical probability, the model created cannot ignore the infinite-collective context of the experiments, and, in the case of games, this set is countable.

384

Let us assume that in a certain game, the player reaches a situation requiring a decision. He or she expects the occurrence of a certain event E, which gives the player the advantage.

The player must choose between a finite number of playing variants $A_1, A_2, ..., A_m$, as result of which event E may occur with probabilities $p_1, p_2, ..., p_m$ respectively.

Assume that $p_1 < p_2 < ... < p_m$.

Denote $p_i = P_{A_i}(E)$, $i = 1, ..., m$ (the probability of the occurrence of expected event E, on condition that the playing variant A_i is chosen, is p_i).

Applying a probability-based strategy means choosing the playing variant A_m, the one offering the player the highest probability of the occurrence of event E.

Why must a player always choose this particular playing variant?

If we answer this question, the choice of the probability-based strategy as optimum is theoretically justified.

Probability theory does not provide any information about the occurrence or nonoccurrence of event E in an isolated gaming situation in which a certain playing variant is chosen, or in a succession of situations in which the playing variants A_i are chosen arbitrarily.

In addition, the law of large numbers cannot be applied in this last case, because we do not deal with experiments performed in identical conditions.

Still, as the only result hinting of the practical aspect of a sequence of experiments, even if the information is one of limit, we must continue to rely on the law of large numbers in the proofs that look for the answer to the optimality problem.

To simplify, we consider the case in which there are only two playing variants A and B (m = 2; the proof can easily be extended to a finite number m of playing variants).

The main condition of the hypothesis is $P_A(E) > P_B(E)$.

We will prove that playing variant A is the optimum choice, in the sense of a definition to be stated later.

The proof uses only the notions of elementary mathematical analysis on R and the law of large numbers.

We identify playing variants A and B through two types of experiments.

To apply the law of large numbers, we need a sequence of independent experiments of same type.

This is why we first consider the countable case of this problem; namely, the case in which the player plays regularly, playing each game with a frequency that allows the creation of a long succession of experiments of type A or B.

Obviously, this is an idealization, because during the player's lifetime, he or she can participate in only a finite number of games.

Still, the idealization cannot be avoided, because the law of large numbers is the only mathematical tool we are allowed to use in the proposed proof.

We refer to this matter again when we consider the finite case.

In the countable case, we call a playing variant (or strategy) a succession of experiments of type A and B, namely, a sequence of experiments (tests) attached to a given gaming situation.

The sequence $\mathcal{V} = (V_1, V_2, ..., V_n, ...)$, with $V_i \in \{A, B\}$, $\forall i \in N^*$ represents a sequence of independent experiments in this way:

V_i is the playing variant (A or B) chosen by a player when reaching the given gaming situation, in experiment number i.

A sequence $\mathcal{V} = (A, B, A, A, B, A, B, A, A, ...)$ means the choice of an order and a number of experiments of type A and B and represents the player's particular strategy chosen for the succession of games (more exactly, for the succession of occurrences of the given gaming situation).

Such a sequence is called a sequence of (A, B) experiments and the whole set of these sequences is denoted by **S**.

For an arbitrary sequence $\mathcal{V}$ of (A, B) type, we make the following denotations:

$v(n)$ = the number of occurrences of expected event E after the first n experiments (the first n terms of the sequence);

n_A = the number of experiments of type A from the first n experiments (the number of terms A from the first n terms of the sequence);

n_B = the number of experiments of type B from the first n experiments (the number of terms B from the first n terms of the sequence);

Obviously, $n = n_A + n_B$.

The ratio $\dfrac{v(n)}{n}$ represents the relative frequency of occurrence of event E within the sequence of experiments $\mathcal{V}$.

Note that this relative frequency cannot be the subject of the law of large numbers because although the sequence consists of independent experiments, they are not of same type.

We are allowed to apply this theorem only in the particular cases in which the sequence contains only experiments of A type or of B type.

We introduce the relations $\geq$, respectively $>$, on the set $\mathbf{S}$, as follows:

Definition: $\mathcal{U} \geq \mathcal{V}$ if and only if exists the limit

$$f = \lim_{n \to \infty} \frac{u(n) - v(n)}{n} \text{ and } f \geq 0.$$

$\mathcal{U} > \mathcal{V}$ if and only if exists the limit $f = \lim\limits_{n \to \infty} \dfrac{u(n) - v(n)}{n}$ and $f > 0$.

It can be easily proved that relation $\geq$ is a (partial) order relation on $\mathbf{S}$:

Assume $\mathcal{U} \geq \mathcal{V}$ şi $\mathcal{V} \geq \mathcal{Z}$. We have:

$$\frac{u(n) - z(n)}{n} = \frac{u(n) - v(n) + v(n) - z(n)}{n} = \frac{u(n) - v(n)}{n} + \frac{v(n) - z(n)}{n}$$

$\mathcal{U} \geq \mathcal{V} \Rightarrow$ exists $\lim\limits_{n \to \infty} \dfrac{u(n) - v(n)}{n} \geq 0$

$\mathcal{V} \geq \mathcal{Z} \Rightarrow$ exists $\lim\limits_{n \to \infty} \dfrac{v(n) - z(n)}{n} \geq 0$

By summing the two inequalities and using the first equality, we obtain that exists $\lim\limits_{n \to \infty} \dfrac{u(n) - z(n)}{n} \geq 0$, that is $\mathcal{U} \geq \mathcal{Z}$.

Thus, transitivity is proved.

An analogue proof may be done for the relation $>$, which is also an order relation on **S**.

On set **S** we introduce the following relation:

Definition: We say that sequence $\mathcal{U}$ is equivalent to sequence $\mathcal{V}$ if and only if exists the limit $f = \lim_{n\to\infty} \dfrac{u(n) - v(n)}{n}$ and $f = 0$.

Denote this by $\mathcal{U} \sim \mathcal{V}$.

It can be easily proved that relation $\sim$ is an equivalence relation:

$$\lim_{n\to\infty} \frac{u(n) - u(n)}{n} = \lim_{n\to\infty} \frac{0}{n} = 0 \ \Rightarrow \ \mathcal{U} \sim \mathcal{U} \text{ (the relation is reflexive)};$$

If $\mathcal{U} \sim \mathcal{V}$, then:

$$\lim_{n\to\infty} \frac{v(n) - u(n)}{n} = -\lim_{n\to\infty} \frac{u(n) - v(n)}{n} = 0 \ \Rightarrow \ \mathcal{V} \sim \mathcal{U} \text{ (the relation is}$$

symmetrical);

If $\mathcal{U} \sim \mathcal{V}$ and $\mathcal{V} \sim \mathcal{Z}$, then:

$$\frac{u(n) - z(n)}{n} = \frac{u(n) - v(n)}{n} + \frac{v(n) - z(n)}{n} \ \Rightarrow \ \text{exists}$$

$$\lim_{n\to\infty} \frac{v(n) - z(n)}{n} = 0 + 0 = 0 \ \Rightarrow \ \mathcal{U} \sim \mathcal{Z} \text{ (the relation is transitive)}.$$

Results that relation $\sim$ is an equivalence relation.

For a sequence $\mathcal{V}$ of (A, B) experiments, we denote by $\hat{\mathcal{V}}$ the set of all sequences $\mathcal{V}'$ from **S** such that $\mathcal{V}' \sim \mathcal{V}$ (the equivalence class of $\mathcal{V}$).

Observation 1: If we rearrange the terms of a sequence $\mathcal{V}$ from **S** in an arbitrary order, the newly obtained sequence still belongs to $\hat{\mathcal{V}}$.

This can be observed immediately by analyzing at limit the expression $\dfrac{v(n) - v'(n)}{n}$ (where $\mathcal{V}'$ is a sequence obtained through a rearrangement of the terms of $\mathcal{V}$): when n tends to infinity, the

388

common terms of the two sequences from the first n terms become increasingly numerous, so the frequency of occurrences of E in the two sequences tends to become equal.

Results that the numerator of the fraction tends to 0, so that fraction tends to $\dfrac{0}{\infty} = 0$.

Observation 2: If $\mathcal{U} \geq \mathcal{V}$ and $\mathcal{V}' \in \hat{\mathcal{V}}$, then $\mathcal{U} \geq \mathcal{V}'$.
The sentence is also valid for the relation $>$.

We can observe that $\dfrac{u(n) - v'(n)}{n} = \dfrac{u(n) - v(n)}{n} + \dfrac{v(n) - v'(n)}{n}$.

According to the hypothesis, exists the limit $\lim\limits_{n \to \infty} \dfrac{u(n) - v(n)}{n} \geq 0$

and exists the limit $\lim\limits_{n \to \infty} \dfrac{v(n) - v'(n)}{n} = 0$; therefore, exists the limit

$\lim\limits_{n \to \infty} \dfrac{u(n) - v'(n)}{n} \geq 0$, that is $\mathcal{U} \geq \mathcal{V}'$.

We show further that in set **S** exists a maximal element (with respect to order relation $\geq$), namely, the sequence
$\mathcal{A} = (A, A, A, A, A, ...)$.
This is the particular sequence containing only experiments of type A.
We want to prove that $\mathcal{A} \geq \mathcal{V}$, for any $\mathcal{V} \in$ **S**.
We consider three cases for the content of sequence $\mathcal{V}$:

1) The number of terms of type A of $\mathcal{V}$ is finite, and the number of terms of type B is infinite.

Denote $f_n = \dfrac{a(n) - v(n)}{n}$.

We have: $f_n = \dfrac{a(n)}{n} - \dfrac{v(n)}{n} = \dfrac{a(n)}{n} - \dfrac{a(n_A) + b(n_B)}{n_A + n_B}$.

(n_A is the number of experiments of type A and n_B the number of experiments of type B from the first n terms of sequence $\mathcal{V}$.)

In the last fraction, we force a reduction by n_B, which is non-zero, and we obtain:

$$(1) \quad f_n = \frac{a(n)}{n} - \frac{\dfrac{a(n_A)}{n_B} + \dfrac{b(n_B)}{n_B}}{\dfrac{n_A}{n_B} + 1}.$$

Observe that when $n \to \infty$, then $n_B \to \infty$ and we can apply the law of large numbers for the relative frequency $\dfrac{b(n_B)}{n_B}$, which corresponds to a sequence of independent experiments of type B, as well for the relative frequency $\dfrac{a(n)}{n}$, which corresponds to a sequence of independent experiments of type A.

We have then $\lim\limits_{n \to \infty} \dfrac{b(n_B)}{n_B} = P_B(E)$ and $\lim\limits_{n \to \infty} \dfrac{a(n)}{n} = P_A(E)$.

When $n \to \infty$, n_A remains constant from a certain rank n upward, as well as $a(n_A)$, so we have $\lim\limits_{n \to \infty} \dfrac{n_A}{n_B} = \lim\limits_{n \to \infty} \dfrac{a(n_A)}{n_B} = 0$.

Using these limits in relation (1), it follows that exists

$$\lim\limits_{n \to \infty} f_n = P_A(E) - \frac{0 + P_B(E)}{0 + 1} = P_A(E) - P_B(E) > 0.$$

This is equivalent to $\mathcal{A} > \mathcal{V}$.

2) The number of terms of type A of $\mathcal{V}$ is infinite, and the number of terms of type B is finite.

With the same denotations as in case 1), we have:

$$f_n = \frac{a(n)}{n} - \frac{a(n_A) + b(n_B)}{n_A + n_B}.$$

In the last fraction. we force a reduction by n_A, which is non-zero, and we obtain:

390

(2) $f_n = \dfrac{a(n)}{n} - \dfrac{\dfrac{a(n_A)}{n_A} + \dfrac{b(n_B)}{n_A}}{1 + \dfrac{n_B}{n_A}}$.

We have $\lim\limits_{n\to\infty} \dfrac{a(n)}{n} = P_A(E)$, $\lim\limits_{n\to\infty} \dfrac{a(n_A)}{n_A} = P_A(E)$ (the law of large

numbers), $\lim\limits_{n\to\infty} \dfrac{n_B}{n_A} = \lim\limits_{n\to\infty} \dfrac{b(n_B)}{n_A} = 0$ (when $n \to \infty$, then $n_A \to \infty$,

while n_B and $b(n_B)$ remain constant from a certain rank n upward).

Using these limits in relation (2), it follows that exists

$\lim\limits_{n\to\infty} f_n = P_A(E) - \dfrac{P_A(E) + 0}{1 + 0} = P_A(E) - P_A(E) = 0$.

This is equivalent to $\mathcal{A} \geq \mathcal{V}$, particularly $\mathcal{A} \sim \mathcal{V}$.

3) The number of terms of type A of $\mathcal{V}$ is infinite, and the number of terms of type B is still infinite.

Using the same denotations as in previous cases, we have:

$f_n = \dfrac{a(n)}{n} - \dfrac{a(n_A) + b(n_B)}{n_A + n_B}$.

In the last fraction, we force a reduction by $n_A n_B$, which is non-zero:

$f_n = \dfrac{a(n)}{n} - \dfrac{\dfrac{a(n_A)}{n_A}\cdot\dfrac{1}{n_B} + \dfrac{b(n_B)}{n_B}\cdot\dfrac{1}{n_A}}{\dfrac{1}{n_B} + \dfrac{1}{n_A}} = \dfrac{a(n)}{n} - \dfrac{\dfrac{a(n_A)}{n_A}\cdot\beta_n + \dfrac{b(n_B)}{n_B}\cdot\alpha_n}{\alpha_n + \beta_n}$

where $\alpha_n = \dfrac{1}{n_A}$ and $\beta_n = \dfrac{1}{n_B}$.

Observe that $\alpha_n, \beta_n > 0$ and $\lim\limits_{n\to\infty}\alpha_n = \lim\limits_{n\to\infty}\beta_n = 0$.

We have: $f_n = \dfrac{a(n)}{n} - \dfrac{a(n_A)}{n_A}\cdot\dfrac{\beta_n}{\alpha_n + \beta_n} - \dfrac{b(n_B)}{n_B}\cdot\dfrac{\alpha_n}{\alpha_n + \beta_n}$.

Let us denote $\varepsilon_n = \dfrac{\beta_n}{\alpha_n + \beta_n}$, $\mu_n = \dfrac{\alpha_n}{\alpha_n + \beta_n}$ and observe that

391

$0 < \varepsilon_n, \mu_n < 1$ and $\varepsilon_n + \mu_n = 1$, for any n.

We have: $\qquad f_n = \dfrac{a(n)}{n} - \dfrac{a(n_A)}{n_A} \cdot \varepsilon_n - \dfrac{b(n_B)}{n_B} \cdot \mu_n =$

$$= \dfrac{a(n)}{n} - \dfrac{a(n_A)}{n_A} \cdot \varepsilon_n - \dfrac{b(n_B)}{n_B} \cdot (1 - \varepsilon_n).$$

If we then add and subtract $\dfrac{a(n_A)}{n_A}$ in the right-hand member, we

obtain:

$$f_n = \dfrac{a(n)}{n} - \dfrac{a(n_A)}{n_A} + \dfrac{a(n_A)}{n_A} \cdot (1 - \varepsilon_n) - \dfrac{b(n_B)}{n_B} \cdot (1 - \varepsilon_n) =$$

$$= \dfrac{a(n)}{n} - \dfrac{a(n_A)}{n_A} + (1 - \varepsilon_n)\left[\dfrac{a(n_A)}{n_A} - \dfrac{b(n_B)}{n_B} \right].$$

We know that exist the limits: $\displaystyle \lim_{n \to \infty} \dfrac{a(n)}{n} = \lim_{n \to \infty} \dfrac{a(n_A)}{n_A} = P_A(E)$

and $\displaystyle \lim_{n \to \infty} \dfrac{b(n_B)}{n_B} = P_B(E)$ (from the law of large numbers).

If sequence ε_n is convergent, the problem is solved because it immediately results that sequence f_n is convergent and its limit is:

$$\lim_{n \to \infty} f_n = P_A(E) - P_A(E) + (1 - \lim_{n \to \infty} \varepsilon_n)[P_A(E) - P_B(E)] =$$

$$= (1 - \lim_{n \to \infty} \varepsilon_n)[P_A(E) - P_B(E)], \text{ which is non-negative because}$$

$P_A(E) > P_B(E)$, and all terms of sequence ε_n are in the interval $(0, 1)$ (so its limit is in the interval $[0, 1]$).

However, in general, sequence ε_n is not convergent.

All we know about this sequence is that is bounded, having its terms in $(0, 1)$.

We have $\varepsilon_n = \dfrac{\beta_n}{\alpha_n + \beta_n} = \dfrac{\dfrac{1}{n_B}}{\dfrac{1}{n_A} + \dfrac{1}{n_B}} = \dfrac{n_A}{n}$, so the terms of this

sequence depend exclusively on the order and number of experiments of type A and type B within sequence $\mathcal{V}$, and these can be of any complexity.

Can we rearrange the terms of sequence $\mathcal{V}$ such that sequence ε_n (corresponding to the new sequence of experiments) is convergent?

If the answer to this question is positive, then we could run the previous proof for the sequence $\mathcal{V}''$ obtained through the rearrangement of terms of $\mathcal{V}$, and get the result that exists $\lim\limits_{n\to\infty} f'_n \geq 0$.

Then, by using Observation 1 and Observation 2, it results that $\mathcal{A} \geq \mathcal{V}'' \sim \mathcal{V}$, therefore $\mathcal{A} \geq \mathcal{V}$.

Well, the terms of sequence $\mathcal{V}$ can be rearranged such that sequence ε_n becomes convergent:

Let us analyze the monotony of sequence ε_n:

$$\varepsilon_n - \varepsilon_{n+1} = \frac{n_A}{n} - \frac{n_A + k}{n+1} = \frac{n_A - nk}{n(n+1)}, \text{ where } k \in \{0, 1\}, \text{ as the}$$

$n + 1$ –th term of the sequence of experiments is of type B ($k = 0$) or of type A ($k = 1$).

We can observe that for $k = 0$ we have $\varepsilon_n - \varepsilon_{n+1} > 0$ and for $k = 1$ we have $\varepsilon_n - \varepsilon_{n+1} < 0$, so the sequence ε_n is not monotonic.

On the other hand, it is monotonic on successive intervals: it is strictly increasing on any interval of consecutive experiments of type A, and it is strictly decreasing on any interval of consecutive experiments of type B.

We want to build a sequence $\mathcal{V}'$, by using the terms of sequence $\mathcal{V}$, such that the newly obtained sequence ε_n is convergent (we use the same denotations as for sequence $\mathcal{V}$).

The construction we want to make is a particular one and is not the only one that ensures the convergence of ε_n.

We choose the first three groups (intervals of consecutive experiments of the same type) of experiments of type A and type B (no matter their number): ABBAA

The first four terms of sequence ε_n are: 1/1, 1/2, 1/3, 2/4.

We complete the sequence with experiments of type B until ε_n exceeds the value for the last experiment of the previous group of type B, namely, 1/3: ABBAAB.

The first six terms of the sequence ε_n are now: 1/1, 1/2, 1/3, 2/4, 3/5, 3/6.

We continue to complete the sequence with experiments of type A until ε_n exceeds the value for the last experiment of the previous group of type A, namely, 3/5: ABBAABAA.

The first eight terms of the sequence ε_n are now: 1/1, 1/2, 1/3, 2/4, 3/5, 3/6, 4/7, 5/8.

Now we complete the sequence again with experiments of type B until ε_n exceeds the value for the last experiment of the previous group of type B, namely, 3/6: ABBAABAAB.

The first nine terms of sequence ε_n are now: 1/1, 1/2, 1/3, 2/4, 3/5, 3/6, 4/7, 5/8, 5/9.

We continue with completing with a group of type A similarly.

This construction continues to infinity by using the same steps.

The fact that the number of experiments of type A and the number of experiments of type B within sequence $\mathcal{V}$ are infinite ensures the possibility of constructing sequence $\mathcal{V}'$ by using the terms of sequence $\mathcal{V}$.

We obtain the following sequences $\mathcal{V}'$ and the corresponding ε_n :

$\mathcal{V}'$: A B B A A B A A B A A B A A B...
ε_n : 1/1, 1/2, 1/3, 2/4, 3/5, 3/6, 4/7, 5/8, 5/9, 6/10, 7/11, 7/12, 8/13, 9/13, 9/14, ...

The resulting sequence ε_n is convergent and $\mathcal{V}' \sim \mathcal{V}$, according to Observation 1.

This particular construction can be done generally by considering any sequence of experiments that fits the following rule: for any group of type A, the values of $\varepsilon_n = \dfrac{n_A}{n}$ in the terminal points of that group are higher respectively than the values of same expression in the terminal points of the previous group of type A.

This condition, along with the boundedness of ε_n, ensures the convergence of this sequence.

The symbolic plot (assumed continuous) of such a sequence looks like this:

AAA B AAAA BB A BB AAAA BB AAA BB AAAAA BB AAAAA BBB AAAAAA BBBBB AAAAAAA BBBBB AAAAAAA BB AAAAA B...

This plot does not represent the plot of a particular sequence ε_n, but illustrates one way to visualize the convergence.

Observe that the length of the groups of type A or B does not count toward convergence; only the order of the terminal points of these groups influences the convergence.

This allows the construction of sequence $\mathcal{V}''$, which depends on the density of terms of type A and B within the initial sequence $\mathcal{V}$.

Let us return to the proof of case 3) of the initial problem.

We saw that for any sequence $\mathcal{V}$, we can rearrange its terms such that sequence ε_n becomes convergent.

In this case, we have that exists

$$\lim_{n \to \infty} f_n = (1 - \lim_{n \to \infty} \varepsilon_n)[P_A(E) - P_B(E)].$$

We have $0 < \varepsilon_n < 1$ and therefore $\lim_{n \to \infty} \varepsilon_n \leq 1$.

Results that the limit $\lim_{n \to \infty} f_n$ is strictly positive if $\lim_{n \to \infty} \varepsilon_n < 1$ and is zero if $\lim_{n \to \infty} \varepsilon_n = 1$, which as a whole is equivalent with $\mathcal{A} \geq \mathcal{V}$.

Thus, all cases of the content of sequence $\mathcal{V}$ are covered (the case in which the number of terms of type A and the number of terms of type B are both finite does not exist, because $\mathcal{V}$ is a sequence).

In this way, we have proved that there is a maximal element with respect to the initially defined order relation $\geq$ ($\mathcal{A} \geq \mathcal{V}$, $\forall \, \mathcal{V} \in S$).

Let us see how this mathematical model represents the initial problem and answers its question by proving that the probability-based strategy is optimum.

The definition of the relation order created on **S** represents the criterion of comparison between two arbitrary playing variants (strategies).

A variant $\mathcal{U} = $ (ABBAAAA...) is *better* than other variant $\mathcal{V} = $ (BABABAA...) only in the sense of this relation order ($\mathcal{U} \geq \mathcal{V}$):

exists the limit $\lim\limits_{n \to \infty} \dfrac{u(n) - v(n)}{n} \geq 0$.

If this limit is strictly positive, results that a natural number N exists such that $\dfrac{u(n) - v(n)}{n} > 0, \forall n \geq N$, so $u(n) > v(n)$ for $n \geq N$.

In other words, the cumulative number of occurrences of expected event E is higher from a certain rank upward if the playing variant $\mathcal{U}$ is chosen.

We have no other information about this rank N; we only know that it exists.

This is the practical coverage of the model we have created, even though it refers to a situation of limit: the property of a playing variant being better than another is translated into achieving a cumulative number of occurrences of expected event E that is higher from a certain rank of the sequence of experiments.

If the limit from the definition of relation $\geq$ is zero, then the two variants are in the same equivalence class (in the sense of the definition of relation $\sim$).

Translated into practice, this means that the two strategies lead to equalization of the relative frequencies of occurrence of event E, the approximation taking effect at limit.

Because relative frequency implies a practical coverage that is lower than that implied by frequency (the cumulative number of occurrences of E), the mathematical model may be refined by defining a relation order within an equivalence class.

By ordering the equivalent sequences through this relation, the playing variants $\mathcal{U} = $ (BBAB (*finite number of terms of type* B) AAA...(*infinite number of terms of type* A)...) and $\mathcal{V} = $ (BBBABAB (*finite number of terms of type* B) AAA...(*infinite number of terms of type* A)...) is no longer *equally good*.

For this particular example, a possible order would be:

$U > V$ if the number of experiments of type B is higher in V than in U.

The practical coverage of such an order would be: even if equalization of relative frequencies of event E takes effect for U and V at limit, the occurrence of the rank N from which a certain approximation is fulfilled may be delayed as the number of experiments of type B of one of the sequences is increased.

Through the model we created, we proved that the playing variant containing only experiments of type A is optimum:

$A = (AAA...)$.

In most of the cases studied in case 1) and a big part of subcases in case 3), the order between A and an arbitrary variant V is strict $(A > V)$.

This fact accomplishes the practical coverage represented by the cumulative number of occurrences of E:

$$A > V \Leftrightarrow \exists \lim_{n \to \infty} \frac{a(n) - v(n)}{n} > 0 \Rightarrow \text{exists a natural number } N \text{ such}$$

that $a(n) > v(n)$ for any $n \geq N$.

This theoretically justifies the choice to play variant A as the optimum (the best, in the sense of the defined relation order); namely, the variant containing only experiments of type A.

Any alteration of this sequence with experiments of type B reverses the order or keeps the newly obtained sequence in the same equivalence class with the first.

Thus, we found a theoretical motivation (for the countable case) for choosing the strategy containing only experiments of type A (the type that offers the highest probability of the occurrence of event E) as optimum.

At a practical, experimental level, the results of applying a probability-based strategy might contradict the optimum attribute, but this is only about appearance.

Because the model uses limit properties, the existence of rank N from which these properties come into effect is theoretical; this number can occur anywhere toward infinity.

Of course, it is possible that the number of experiments in which a player participates over a given period (even over a lifetime) will never reach this rank N, so the favorable effect of the strategy, even if it exists mathematically, is impossible to achieve in reality.

Still, this effect may appear in practice if the difference $P_A(E) - P_B(E)$ is big enough, because the positive limit of f_n depends on this value, as well as the rank N from which $a(n) > v(n)$.

In fact, this reverts to the stabilization of frequencies at the experimental level, about which we talked in the section dedicated to the philosophy of probability: there exists *something* that makes the frequency of occurrences of an event become stabilized after a certain number of independent experiments performed in the same conditions, and this stabilization is much more visible (in effect, the number of experiments through which it becomes visible is lower) as the probability of that event is higher.

It is possible (but improbable) to get no heads (that is, all tails) after 100 throws of a coin, although the probability of this event is 1/2. This is the expression itself of the conceptual distinction between *possible* and *probable*.

Still, in practice such a thing never seems to happen; the relative frequency of the occurrence of heads oscillates around 1/2 from a rank of experiments much lower than 100.

This also happens in the case of the favorable effect of the probability-based strategy: the bigger the difference in probabilities $P_A(E) - P_B(E)$, the lower the rank of experiments from which the difference in frequencies $a(n) - v(n)$ becomes positive (visible).

The proof of optimality was performed for the particular case in which only two playing variants A and B exist.

Both the mathematical model and the proof itself can be extended to a decisional situation having a finite number of variants $A_1, A_2, ..., A_m$.

The model created and the proof presented give us theoretical motivation for choosing the probability-based strategy as the optimum strategy in the countable case.

How do we transfer these results to the finite or even isolated case (of a single experiment)?

How do we motivate the choice of experiment of type A or of a finite group of type A in the case—obviously, practical—in which a player participates in only a limited number of experiments?

Returning to a suggestive example of a decisional situation from the introductory chapter, how do we theoretically motivate the person in the phone booth to consider calling the house with more people living in it the optimum choice?

Well, a mathematical model with which to represent this problem exclusively or any similar finite case does not exist because, as we have said several times before, infinity is part of the definition of probability and the probability-based strategy includes probability as a criterion.

To extend the motivation of optimality from the countable case to the finite case, we must include the finite case in a hypothetical collective context.

The following might stand for a theoretical motivation: a player must choose to play variant A (the one offering the highest probability of the occurrence of an expected event), except for the fact that, in the case of a sequence of experiments (in similar gaming situations) continuing (to infinity), this sequence must not be altered by including experiments of type B because it would no longer be optimum (in the sense of the relation order defined in the countable case).

Obviously, such a motivation has little practical coverage, in fact, it even weakens the practical coverage of the countable case: we already know that the sequence of experiments from the countable case is hypothetical (it cannot be practically fulfilled); now we add another hypothetical situation, the prolongation of the finite group of experiments.

Still, at a purely theoretical level, this is the only motivation of the optimality of probability-based strategy.

Even if this motivation is enhanced and refined, it cannot ignore the *infinite* element because the concept of probability is still involved.

Also in the finite case, the situations in which the difference $P_A(E) - P_B(E)$ is big enough may accomplish the practical coverage of the theoretical motivation: as this difference increases, the rank N of experiments from which the favorable effect of the choice is fulfilled may diminish until it reaches the finite group of performed experiments.

As we mentioned earlier, the motivation provided by this mathematical model is eminently theoretical.

In reality, a player build his or her own playing strategy by factoring in several criteria, including subjective ones, and many times its application has better practical results than the probability-based strategy does.

This happens not only in games but also in the entire decisional ensemble of daily life.

The term *strategy* contains in its essence both practical and theoretical components.

The theoretical motivation of optimality of a certain strategy does not ignore the practical component, but this component is in fact a *theoretically practical* one within the mathematical model.

This delimitation is another expression of the conceptual distinction between mathematical probability and the subjective interpretations of probability.

VARIOUS APPLICATIONS

We now propose to solve several probability calculus applications for finite cases. Their difficulty grows progressively from simple to the intermediate level.

The problems presented are addressed to readers who went through the *Beginner's Calculus Guide* and the solved applications in previous chapters and want to improve their problem-solving skills.

1) In a student's briefcase are seven math books, two English books and two drawing books. What is the probability that a randomly chosen book is a math book?

2) Five people, A, B, C, D and E, want to go to a concert, but they have only three tickets. They decide to randomly select who is to go. What is the probability that:
  a) A is selected?
  b) B is not selected?
  c) C and D are selected?

3) Each of first seven natural numbers is written on a label. The labels are mixed up, then placed side by side. What is the probability of the number 4 coming right after the number 3? What is the probability that the seven numbers are placed in increasing order?

4) If boys and girls are born equally likely, what is the probability that in a family with three children, exactly one is a girl?

5) Find the probability that, writing randomly a number of three distinct digits, this number is a perfect square.

6) What is the probability that, upon opening a 75-page book at random, we see a page numbered with a prime number (assuming the first page is numbered 1)? What is the probability that we see a perfect square number?

7) Given the sets A = {1, 2, 3, 4, 5, 6, 7} and B = {2, 3, 5, 7, 8}, find the probability that, choosing at random one number from each set, the two numbers are equal.

8) A coin is thrown three times. What is the probability of:
a) heads coming up exactly two times;
b) heads coming up at least once;
c) heads not coming up two times consecutively.

9) Two dice are rolled. Find the probability that:
a) the sum of the rolled numbers is 6;
b) the sum of the rolled numbers is less than 6;
c) two even numbers are rolled;
d) number 6 comes up on at least one die;
e) the first die shows a number greater than the second.

10) A 7-digit number is obtained by arranging the figures 1, 2, 3, 4, 5, 6, 7 at random. What is the probability that the obtained number:
a) has 2 as the last digit;
b) is a multiple of 5;
c) is a multiple of 3.

11) Twelve people are randomly seated at a round table. What is the probability that a particular couple sit next to each other?

12) A soccer match may have these outcomes: hosts win (1), visitors win (2) or draw (X). A person predicts the outcomes of four matches by writing 1, 2, X, at random. Assuming the outcomes of each match are equally probable, what is the probability of:
a) all four predictions being correct?
b) none of predictions being correct?
c) exactly one prediction being correct?

13) Seven persons are in an elevator. The elevator stops on ten floors. What is the probability that no two persons exit on the same floor?

14) Find the probability that among k randomly selected digits:
a) there is no digit 0;
b) there is no digit 2;
c) there is no digit 0 or there is no digit 2.

15) n persons, A and B among them, form a rank in arbitrary order. What is the probability that there are exactly r persons between A and B?

16) A rod of length r is broken at random into two parts.
What is the probability that the length of the shortest part does not exceed $5r/6$?

17) A person A receives information and transmits it to another person B. Person B transmits it to a third person C, who transmits it to a fourth person D.
Knowing that each person tells the truth in one case out of three, find the probability that A told the truth, given that D told the truth.

18) At a game of American roulette (which uses 38 numbers), a player places chips of same value on several singular numbers.
How many numbers must the player bet on so that the probability of winning (no matter the amount) is greater than 1/7?

19) At a blackjack game with a 52-card deck, find the probability of a player:
a) achieving at least 15 points from a maximum of three cards;
b) achieving a maximum of 20 points from a maximum of five cards;
c) exceeding 21 points from a maximum of five cards.
(Calculate these probabilities with the understanding that there are no other cards showing.)

20) A single player participates in a blackjack game with a 52-card deck. The player receives 5, J, 2 and stops at 17 points.

The dealer is in play now, and turns a Q.

Find the probability (calculated in this gaming moment) that:

a) the dealer beats the player at the next card;

b) the dealer finally beats the player;

c) the dealer finally busts (exceeds 21 points).

21) Solve problems 19) and 20) for a game with two 52-card decks.

22) Horses A, B and C participate in a race, and are ridden by the jockeys a, b and c.

A jockey is the owner of the horse with the same letter. The jockeys are chosen at random to ride the horses.

What is the probability that a horse is ridden by its owner?

Generalize this for n horses and n jockeys.

23) Thirty books are randomly archived on three shelves.

What is the probability of a shelf (no matter which) containing ten books?

Generalize this for arbitrary numbers a – number of books, b – number of shelves and c – number of books on a shelf ($c < a$).

24) On a segment of length r two points are taken at random.

What is the probability that the distance between these points does not exceed kr, where $0 < k < 1$?

25) A regular n–gon is inscribed into a circle of radius R.

A point is thrown into the circle. What is the probability that the point falls inside the n–gon?

26) Two vessels must arrive at the same wharf. Ship arrivals are independent random events, equally probable over a 24-hour period.

Find the probability that one of the vessels must wait until the wharf is vacant, if the berthing time for the first vessel is one hour and for the second is two hours.

27) At a game of European roulette (which uses 37 numbers), a player bets amount S on the color red and amount $S/2$ on each of n black numbers.

What values can n take such that the probability of winning (no matter the amount) is greater than 1/5?

Generalize this for the arbitrary numbers: S – the amount bet on red, S/m – the amount bet on each of the n black numbers, where S is positive, m and n are natural numbers, $m > 1$, $1 < n < 38$.

28) Among $t = 50$ lottery tickets, $w = 17$ win prizes.
We buy $b = 7$. What is the probability that $g = 3$ wins?
Generalize this to arbitrary numbers t, w, b, g.

29) Three shooters fire at a target. The probability of hitting the target is p, q, r, for each shooter, respectively.

After the shooting ends, the shooters find that the target was hit just once.

What is the probability that the target was hit by the first shooter?

30) For the first stage of the mathematics Olympiad, twenty algebra and fifteen geometry problems must be solved.

From among them, two algebra and two geometry problems are to be solved in a written test.

The pupil who solves three of the four problems qualifies for the superior stage of the competition.

One pupil solved seventeen algebra and eleven geometry problems until the test date.

What is the probability for that pupil to qualify (assuming that a previously unsolved problem cannot be solved during the test)?

31) A student must sit for an examination consisting of four questions selected randomly from a list of 150 questions.

To pass, the student must answer all four questions.

What is the probability that the student passes the examination if he or she knows the answers to 100 questions on the list?

Generalize this.

32) n shooters simultaneously fire at a mobile target.

The probability of hitting the target is the same for all shooters and is equal to $1/k$, where k is a non-negative natural number.

Calculate the probability that the target is hit by at least one shooter.

33) From an urn that contains n white and m black balls, k balls are drawn at random. What is the probability that there are r ($r \leq n$) white balls among them?

34) We have four urns; the first contains five white and four black balls, the second contains three white and six black balls, the third contains two white and five black balls and the fourth contains two white and three black balls.

A ball is drawn from a randomly chosen urn. Find the probability of the drawn ball being black.

35) We have four urns with the same composition as in problem 34). Two balls are drawn from two randomly chosen urns (one ball from each). Find the probability of:

a) drawing two white balls;
b) drawing two black balls;
c) drawing two balls of different colors.

36) We have an urn with the following composition: three white, four green, five yellow and two red balls. Ten draws are performed, and the drawn balls are returned to the urn after each draw.

Calculate the probability of obtaining from the ten draws:

a) three white, three green, two yellow and two red balls;
b) only yellow balls;
c) balls of exactly two colors;
d) balls of all colors (four).

37) We have $n - 1$ urns. The first urn contains a white ball and $n - 1$ black balls, the second contains two white and $n - 2$ black balls, ..., the $n - 1$ -th urn contains $n - 1$ white balls and a black ball.

One ball is randomly drawn from each urn. Calculate the probability that at least one of the drawn balls is white.

38) There are seven white balls and two black balls in an urn. n balls are randomly drawn.

Determine n, such that the probability of being at least a black ball among them is greater than 1/2.

39) Half of the population of a town are people of type A and the other half are people of type B. A person of type A never says the truth and a person of type B tells the truth with a probability of 2/5.

What is the probability that a randomly selected citizen gives a true answer to a question?

40) At an exam, the probabilities that three candidates solve a certain problem are 4/5, 3/4, and 2/3, respectively.

Calculate the probability for the examiner to receive from these candidates:

a) exactly one correct solution;

b) at most one correct solution;

c) at least one correct solution.

41) A group of fifteen homing pigeons is sent to a destination. The probability of a pigeon returning home is $p = 0.4$ (the same for all pigeons). Find the probability that:

a) ten pigeons come back; generalize this for n pigeons, from which m come back;

b) at least seven pigeons come back; generalize this for n pigeons, from which at least m come back;

c) at most twelve pigeons come back; generalize this for n pigeons, from which at most m come back $(m < n)$.

42) The probability that a goalkeeper of a soccer team blocks a penalty ball is $p = 0.3$. Find:

a) the probability that a goalkeeper blocks four penalty balls;

b) the probability that a goalkeeper takes at most two goals from three penalty kicks;

c) the highly probable number of goals when four penalty kicks are executed.

43) A die is rolled six times. Calculate the probability that all six faces are shown. Generalize this for n rolls.

44) Twelve dice are rolled. What is the probability of twice obtaining each of the numbers 1, 2, 3, 4, 5, 6 ?

45) Three distinct vertexes of a regular $2n + 1$ –gon are randomly chosen. What is the probability that the center of the $2n + 1$ –gon is inside the triangle determined by the three chosen points?

46) Five dice are rolled. Find the probability that:
a) the sum of the numbers rolled is 25;
b) the sum of the numbers rolled is greater than 17;
c) the numbers rolled are different;
d) a triple and a double are rolled;
e) two doubles are rolled.

47) A die is rolled twenty times. Find the probability that:
a) number 5 is rolled exactly four times;
b) number 5 is rolled at least five times;
c) an even number is rolled exactly ten times;
d) the same number is shown on two consecutive rolls (at least one pair of consecutive rolls).

48) For a slot machine with five reels and eight symbols, find the probability of obtaining:
a) a specific symbol on exactly one reel (no matter which reel);
b) a specific symbol on exactly two reels (no matter which reels);
c) five different symbols.

49) There are r people in a group. What is the probability that at least two of them were born in the same month?

50) Calculate the probability that the birthdays of twelve individuals are in different months.

51) Given a group of thirty people, calculate the probability that two birthdays fall on each of six months of the year and three different birthdays are each of the remaining six months.

52) Five cards are simultaneously drawn at random from a 52-card deck.

Calculate the probability that:
a) the five cards contain at least two queens;
b) the five cards contain at least three hearts;
c) the five cards contain at least three suited cards;
d) the five cards contain all the four symbols.

53) Let p_x^t be the probability that a person of age x years will still be alive after t years.

Show that $p_x^t = p_x^1 \cdot p_{x+1}^1 \cdot \ldots \cdot p_{x+t-1}^1$.

54) Five dice are rolled one after another. What is the probability that:
a) the sum of the points shown on the first two dice is greater than that shown on the last three dice?
b) there are two dice such that the sum of points shown by these dice is greater than that shown by the other three dice?

55) Eight persons, four women and four men, are randomly split in two equal groups.

What is the probability that each group has the same number of women and men?

Generalize this for a group of $2n$ persons, from which n are women and n are men.

56) A coin of diameter d is thrown onto a parquet floor.

The parquet has the form of squares with side a, $a > d$.

What is the probability that the coin does not intersect any of the sides of parquet squares?

57) A poll company publishes a survey in three newspapers A, B and C, which have readers in proportion 2, 3, 1.

The probabilities that one reader answers the survey are 0.002, 0.001 and 0.0005, respectively.

a) If the company receives an answer, what is the probability that the sender is a reader of newspaper A? How about newspaper B? How about C?

b) If the company receives two answers, what is the probability that both senders are readers of newspaper A?

(We assume that each reader reads a single newspaper.)

58) A man goes to work every day with his car and may be delayed at two intersections, A and B.

The probability of being delayed at intersection A is 0.8, at intersection B is 0.5 and these delays are independent of one another.

The man is late for work only if he is delayed at both intersections.

Given that the man is late for work one day, what is the probability that he was delayed at intersection A?

59) A bus line has three stops. The probabilities of delaying at these stops are 0.3, 0.5 and 0.7.

a) Find the probability of being no delay at the three stops.

b) Find the probability of being exactly one delay.

c) Given that is one single delay, find the probability for that delay to be at the first stop.

60) A device containing N lamps becomes out of order when a lamp becomes defective.

The probability of a lamp becoming defective is p (the same for all lamps) and the lamps become defective independently of each other.

Knowing that the device is out of order and that detecting the defective lamp is through successively replacing each lamp with a new one, find the probability of detecting the defective lamp by making n replacements.

61) The probability of occurrence of an event A in each test is 1/2.

Calculate the probability that the number of occurrences of event A in 150 independent tests is between forty-five and sixty.

Generalize this for arbitrary numbers: p – probability of occurrence of A in each test, n – number of independent tests, a – minimal number of occurrences, b – maximal number of occurrences.

62) You are participating in a classical poker game with 32 cards and you are dealt (7♠ 9♠ J♠ J♦ Q♥).

Calculate the probability that:

a) a specific opponent holds at least one J;

b) you receive two spades if you discard (J♦ Q♥);

c) you receive at least one J if you discard (7♠ 9♠ Q♥);

d) you receive J or Q if you discard (7♠ 9♠);

e) you receive J and Q if you discard (7♠ 9♠);

f) you achieve a straight if you discard (7♠ J).

63) You are participating in a Texas Hold'em Poker game with n opponents and you are dealt (A♥5♥).

Calculate the probability of:

a) none of your opponents holding A;

b) none of your opponents holding K;

c) none of your opponents holding A or 5;

d) at least one of your opponents holding two spades;

e) at least one of your opponents holding two spades or two clubs or two diamonds;

f) none of your opponents holding two hearts;

g) the flop cards containing at least two hearts;

h) the flop cards containing at least one A;

i) the community cards containing at least two aces after the river;

j) the community cards containing at least three hearts.

(Calculate these probabilities in the moment right before the flop.)

64) You are participating in a Texas Hold'em Poker game with n opponents, you are dealt (4♣5♣), and the flop comes with (3♥J♥K♣).

Calculate the probability that:

a) you achieve a straight by the river;

b) you achieve a flush by the river;

c) at least one opponent holds a J or K;

d) at least one opponent holds two hearts.

(Calculate these probabilities in the moment right before the turn.)

65) (De Mere's paradox) Prove that to obtain at least one 1 at a throw of four dice is more probable than to obtain, at least once, two 1's at twenty-four throws of two dice.

66) n points divide a circle into equal circular arcs.

Two points are randomly chosen from them. What is the mean of the distance between the chosen points?

67) (Buffon's problem) Parallel straight lines are drawn on the plane at a distance $2a$ from one another.

A needle of length $2r$ (r < a) is thrown at random onto the plane.

What is the probability that the needle intersects one of these straight lines?

References

Bărboianu, C., *Probability Guide of Gambling*. Infarom, Craiova, 2003.

Bărboianu, C., *Texas Hold'em Odds*. Infarom, Craiova, 2004.

Bărboianu, C., *The Probability-based Strategy*. Gambling Tribune, 2003.

Boll, M., *Certitudinile hazardului* (*Certainties of the hazard*). Editura ştiinţifică şi enciclopedică, Bucharest, 1978.

Borel, E., *Àpropos of a Treatise of Probabilities*, in the volume *Studies in Subjective Probability*. Academic Press, New York, 1965.

Crăciun, C.V., *Analiză reală* (*Real Analysis*), course of Measure Theory at University of Bucharest, 1988.

Cuculescu, I., *Teoria probabilităţilor* (*Probability Theory*). Bic All, Bucharest, 1998.

Kneale, W., *Probability and induction*. Cambridge, 1949.

Laplace, P., *Essai philosophique sur les probabilites* (*Philosophical Essay on Probabilities*). Paris, 1920.

Onicescu, O., *Principiile teoriei probabilităţilor* (*The Principles of Probability Theory*). Editura Academiei, Bucharest, 1969

Onicescu, O., *Principii de cunoaştere ştiinţifică* (*Principles of Scientific Research*). Oficiul de Librarie, Bucharest.

Trandafir, R., *Introducere în teoria probabilităţilor* (*Introduction to Probability Theory*). Albatros, Bucharest, 1979.

Printed in the United States
58308LVS00003B/32